T0320182

Decision Options®

The Art and Science
of Making Decisions

Series Editor

Michael K. Ong

Stuart School of Business

Illinois State of Technology

Chicago, Illinois, U. S. A.

Aims and Scopes

As the vast field of finance continues to rapidly expand, it becomes increasingly important to present the latest research and applications to academics, practitioners, and students in the field.

An active and timely forum for both traditional and modern developments in the financial sector, this finance series aims to promote the whole spectrum of traditional and classic disciplines in banking and money, general finance and investments (economics, econometrics, corporate finance and valuation, treasury management, and asset and liability management), mergers and acquisitions, insurance, tax and accounting, and compliance and regulatory issues. The series also captures new and modern developments in risk management (market risk, credit risk, operational risk, capital attribution, and liquidity risk), behavioral finance, trading and financial markets innovations, financial engineering, alternative investments and the hedge funds industry, and financial crisis management.

The series will consider a broad range of textbooks, reference works, and handbooks that appeal to academics, practitioners, and students. The inclusion of numerical code and concrete real-world case studies is highly encouraged.

Published Titles

Decision Options®: The Art and Science of Making Decisions, **Gill Eapen**

Emerging Markets: Performance, Analysis, and Innovation, **Greg N. Gregoriou**

Introduction to Financial Models for Management and Planning, **James R. Morris and John P. Daley**

Stock Market Volatility, **Greg N. Gregoriou**

Forthcoming Titles

Portfolio Optimization, **Michael J. Best**

Proposals for the series should be submitted to the series editor above or directly to:
CRC Press, Taylor & Francis Group
4th, Floor, Albert House
1-4 Singer Street
London EC2A 4BQ
UK

CHAPMAN & HALL/CRC FINANCE SERIES

Decision Options®

The Art and Science of Making Decisions

Gill Eapen

Decision Options, LLC
Groton, Connecticut, U. S. A.

CRC Press
Taylor & Francis Group
Boca Raton   London   New York

CRC Press is an imprint of the
Taylor & Francis Group, an **informa** business

A CHAPMAN & HALL BOOK

Chapman & Hall/CRC
Taylor & Francis Group
6000 Broken Sound Parkway NW, Suite 300
Boca Raton, FL 33487-2742

© 2009 by Taylor and Francis Group, LLC
Chapman & Hall/CRC is an imprint of Taylor & Francis Group, an Informa business

No claim to original U.S. Government works

Printed in the United States of America on acid-free paper
10 9 8 7 6 5 4 3 2 1

International Standard Book Number: 978-1-4200-8682-9 (Hardback)

Library of Congress Cataloging-in-Publication Data

Eapen, Gill.
   Decision options : the art and science of making decisions / Gill Eapen.
      p. cm. -- (Chapman & Hall/CRC finance series)
   Includes bibliographical references and index.
   ISBN 978-1-4200-8682-9 (hardcover : alk. paper)
    1. Investments--Decision making. 2. Investment analysis. 3. Finance--Decision making. I. Title. II. Series.

HG4515.E27 2009
658.15'2--dc22
                                                2009013042

Visit the Taylor & Francis Web site at
http://www.taylorandfrancis.com

and the CRC Press Web site at
http://www.crcpress.com

Contents

Preface

This book is a culmination of my experiences in a variety of companies I have worked in or consulted with during the past two decades, spread around life sciences, energy, technology, financial services, and other areas. In most cases, my focus has been financial analysis and planning related to investment decisions—project selection, design, portfolio management, mergers, and acquisitions. Often, I found significant effort expended in analysis—to precisely define cash flows and timings—as well as identifying proxies (similar transactions in the past) in an effort to determine an appropriate discount rate to calculate a net present value (NPV). I also found that scenario analysis is prevalent: to get a better feel for pessimistic and optimistic outcomes in an effort to improve decision quality.

However, when it came down to making a decision, much of the analysis was ignored. The eventual decision was largely based on qualitative aspects or "gut feel." In some cases, the financial analysis was modified to support the decision made—either by changes in projected cash flows or the discount rate used. This led me to wonder why companies spend so much effort in financial analysis up front if the results of such analysis do not aid the decision. It occurred to me that discounted cash flow (DCF) analysis may not be fully capturing the rich details underlying the decision, and the responsible decision maker incorporates these details through qualitative considerations. The decision maker may be using the cash flow analysis to ensure that the assumptions are debated among various stakeholders, and that he or she is not missing anything "big." The result of the DCF analysis itself may not be that relevant for the decision.

This pattern of financial analysis and decision making has been consistent in all industries to which I have been exposed. This has led me to believe that there is something inherently absent in the traditional methodologies that makes them less useful for decisions. Missing are two important considerations: uncertainty and flexibility. Both of these attributes drive the value of an investment and are not systematically considered in traditional financial analysis. Since the decision maker intuitively knows this, he or she will attempt to incorporate these on top of a somewhat rigid NPV in an attempt to make the right decision and enhance the value of the company.

This led me to create a tool set that allows the incorporation of uncertainty and flexibility in investment decisions up front, but not to make it too complex so it defies explanation and inhibits effective communication. In doing so, we eliminate uncertainty in outputs (scenario analysis) but capture uncertainty in inputs: costs, timelines, risks, and revenues. Cash flows are not considered precise, thus eliminating significant time spent in data gathering and consensus building. Imprecise descriptions of cash flows are then married with a rich understanding of future flexibility, focusing decision makers and analysts on the assumptions that drive economic value and determine the decision. The result is less time spent in data collection and the creation of complex spreadsheets and more time spent on the assumptions that drive the decisions, including uncertainty and flexibility. It allows the incorporation of the qualitative attributes of the decision problem and the gut feel of the decision maker, leading to an analysis that not only is useful in making the decision but also provides an effective communication tool for consensus building. I hope the reader will find this tool useful in reducing the time and effort needed to reach decisions and building consensus, improving decision fidelity and communication, and enhancing the value of the enterprise.

Author

Gill Eapen is the founder and managing principal of Decision Options, LLC, a boutique advisory services company. Decision Options pioneered the practical application of options-based valuation of private assets to aid decision making in industries that show high levels of uncertainty and flexibility, such as life sciences, energy, technology, and financial and legal services. He conceived and led the development of a technology platform—Decision Options Technology—that allows modeling of complex decision problems and assets to improve decision making, risk management, and portfolio maximization. At Decision Options, Gill also designed and built an automated analysis and trading system—Decision Options Equity Research—for publicly traded securities.

Prior to establishing Decision Options, Gill was group director at Pfizer, responsible for the financial analysis and planning of the research-and-development portfolio with investments exceeding $4 billion per year and economic value exceeding $50 billion. During his tenure at Pfizer, Gill conceived and led the development of a forecasting and capital allocation methodology and system that incorporates uncertainty emanating from all aspects of pharmaceutical research and development. Before that, he was manager at Deloitte Consulting Group, providing advisory services to a variety of clients in software, technology, commodities, consumer goods, logistics, and manufacturing. Previous employers also include Hewlett-Packard Company and Asea Brown Boveri.

Gill holds graduate degrees from the University of Chicago and Northwestern University as well as an undergraduate degree from the Indian Institute of Technology.

1

Introduction

This book is meant for anyone approaching investment decision making systematically and analytically when uncertainty and flexibility are present in areas such as life sciences, commodities, energy, technology, manufacturing, and financial services. Uncertainty typically exists in all aspects of investment decisions—timelines, costs, success rates, market potential—and is often ignored in traditional financial analysis and decision making. *Flexibility* means the ability to select, defer, abandon, expand, switch, and optimize among alternatives that drive current and future decisions. The term *decision options* is used to represent decisions that have option-like characteristics. Readers may be familiar with financial options (such as a call option on a stock). The price of a call option is driven by the uncertainty in the underlying asset (on which the option is based, such as a stock) and the inherent flexibility (the right but not the obligation to exercise) in them. Decision options are decisions in real life and on real assets driven by the uncertainty in the underlying assumptions and the flexibility in the nature and timing of the decision. The value of decision options is driven by the same parameters that influence the value of financial options. However, decision options are not single-standing options and cannot be solved by available closed-form formulas.

A *decision* is a specific action one can take in the present or in the future but not in the past. An *option* is a right but not an obligation to do something in the future. Decision options are generalized representations of a decision that may include financial and real options, cash flows, risks, swaps, switchers, and other constructs that may or may not interact with each other. An option has economic value only if the parameters that drive it have uncertainty and flexibility. Similarly, the decision options framework will reduce to traditional frameworks if either uncertainty or flexibility is absent. If flexibility is absent but uncertainty is present, the issue will reduce to scenario analysis or Monte Carlo simulation. If flexibility is present but uncertainty is absent, it will reduce to decision tree analysis. If neither uncertainty nor flexibility is present, it will reduce to discounted cash flow analysis. Thus, decision options represent a super set of all the traditional techniques—discounted cash flow, decision tree analysis, Monte Carlo simulation—typically deployed for financial analysis and investment decision making.

The book has a singular focus: economic value-based decision making that is normative (market based). The content is useful to analysts, managers, and decision makers involved in the selection, design, risk management, and portfolio optimization of investment opportunities. It is also an introduction to students of business and finance at the graduate level. It can be assigned reading for business courses in the area of strategy, valuation, risk management, and decision making.

Chapters 2 and 3 contain a qualitative introduction to decision options. Those who are well versed in decision theory and real options can skip these and proceed directly to Chapter 4. In Chapter 2, I introduce decisions and options and describe how they are related. Chapters 3, 4, and 5 contain the introduction of decision options and the quantitative treatment of private and market risks. Chapter 3 illustrates how decisions can be viewed as a combination of three fundamental entities: options, predetermined outcomes, and risks. Chapter 4 describes in detail what risks can be characterized as private risks, how they relate to decisions, and how they can be captured and represented in decision problems. Chapter 5 explores market risks and the use of stochastic processes to represent them in decision problems.

Chapter 6 contains an introduction to self-standing options such as financial options and options pricing theory. This chapter touches on risk-neutral pricing of options and the assumptions underlying financial options pricing. It also provides a brief introduction to the Black-Scholes equation, a closed-form solution for single-standing options when the underlying asset follows a specific price process called random walk.

Chapter 7 describes how decision options are different from single-standing financial options and why techniques of financial options pricing do not directly transfer to the decision options world. An introduction to a software tool, Decision Options Technology (DoT), is provided in illustrating how to formulate, model, and solve simple decision options problems. Chapter 8 deals with a special case of decision options in pricing employee stock options, something that has characteristics of both financial and decision options.

Chapters 9–12 deal with case studies of practical applications of decision options. We look at applications in different industries such as pharmaceuticals, energy, manufacturing, and financial services. Chapter 9 deals with life sciences, including pharmaceuticals, biotechnology, and devices. This is followed by technology and manufacturing examples in Chapter 10. Chapter 11 provides examples from the different world of commodities, including publicly traded metals and energy. Finally, Chapter 12 provides an assortment of cases in financial services including venture capital, incentives, and hedging using financial instruments.

In Chapter 13, some of the common misperceptions that exist around the term *real options* and the use of it in decision making are discussed. Chapter 14 deals with the impediments that currently exist in large companies for the systematic practice of decision options. These include academic challenges as well as many practical issues that one has to overcome in the quest to move organizations to consider uncertainty and flexibility in decisions.

Finally in Chapter 15, I provide a qualitative assessment of how organizations have changed and what they have to focus on to become successful in the future.

2

Decisions, Options, and Decision Options

We all make decisions. Sometimes we make them without much thought, and sometimes we spend a lot of time analyzing. For example, the decisions made by a batter in a baseball game are taken rapidly, possibly based on information on the pitch as well as other information such as the style of the pitcher, weather conditions, and the state of the game. On the other hand, the decision made by Intel to site a new manufacturing facility in South Korea was taken after considerable analysis. Such an analysis may have considered market conditions, price and margin expectations, cost characteristics, currency expectations, country risk, technology trends, as well as the possible actions of the competitors and the governments.

Some of this information is not precise; some is no more than a guess. But, imprecise information is always better than no information, and decisions based on imprecise information are better than random ones. It is possible that the idea of "imprecise information" gives many in business an uncomfortable feeling. This is because the existing analytical framework requires deterministic data, and we grew up in a world where imprecision or uncertainty was always considered "bad" and to be avoided in decision making. This led us to analyses that need deterministic and precise data—a single number to represent an input.

However, outside the corporate environment, many are more comfortable using imprecise information in decision making. For example, we almost always make decisions based on imprecise information in our personal lives. The decision to carry an umbrella on a trip is a decision based on available but imprecise information. There are big decisions, such as the decision to move from one job to another, the decision to buy a home, the decision to return to school, the decision to get married, the decision to have children, and the like. There are also less significant decisions such as what to have for dinner, when to exercise, and whether to travel by air or by car. In all cases, we are given some imprecise information, some based on past experience and some based on forecasts. The information gives us an assessment of various types of risks, and it may present us with choices or options. We may intuitively weigh the risks in various choices or perform a valuation of the choices considering costs, benefits, and risks to pick the best possible one. We may pick a choice that appears to be of the highest value and in some cases to be of the lowest risk. In any case, we all employ a "decision process" to reach a conclusion. If there is not enough time to employ a systematic decision process, as in the case of imminent danger, we may fall back on our instincts, a sort of autopilot that takes over decision making, mostly driven by past experiences.

We all have differing decision processes to take imprecise information and reach conclusions. Some of us may be driven more by "intuition," a feeling that one way is right compared to others, and some may be more analytical, with preference for detailed analysis before reaching a decision. There are many different types of intuitive decision makers also; for some, intuition is based on experience accumulated from the better or worse outcomes of decisions made in the past in similar situations. For others, intuition may be merely a feeling uncluttered by experience or historical data. There are many different types of analytical decision makers as well; some may reduce imprecise data to more deterministic figures and use them to reach conclusions. For others, imprecise information allows a probabilistic understanding, and decisions are made by minimizing risk or maximizing value. Some of these analyses are not necessarily mathematical but may depend on other senses, such as decisions made from visualization of the imprecise information. Such decisions may depend on pattern finding and comparing such patterns to past experiences. The brain is a complex and fascinating machine, able to digest and visualize imprecise information, guess optimally, and conclude fast.

The competency in decision making may have deep roots in human evolution. The ability to make good decisions would have been a very important survival skill: how much food to store or consume, where and when to hunt, where and when to sleep, whom to trust and whom not to trust, and more are all decisions that would have made a difference between life and death for *Homo sapiens* as they ventured out of the African savannahs. Those who had a slightly higher probability of getting this right than wrong survived. Over the ages, the selection process may have resulted in certain types of decision processes dominating others.

We cannot, however, assume that the decision processes most employed today are the best for two important reasons. First, selection acts on a bundle of characteristics exhibited by an individual; thus, any single competency cannot be selected optimally. This is one reason we have a variety of diseases that were "selected" along with certain other attributes that were more important at certain other times. Second, the environment and the availability and the type of information that existed tens of thousands of years ago were substantially different from today. So, all we can assume is that how our brains process information and reach conclusions may have an evolutionary base, but what we did may not have been optimal then and is likely not optimal now.

One of the more difficult decisions for us is making judgments about another human being, whether it is in hiring, marriage, or in court. Although many may not admit it, there are certainly emotions generated by visual images such as the shape of the skull or face that get tangled up in other imprecise information as we make decisions. Pattern matching may be at work here, perhaps based on recent experiences or based on evolutionary empiricism. Modern humans have gotten a lot better in recent times at counteracting such forces, and we have "modernized" the decision processes to be more "objective." In some sense, we have better human software that can mask these tendencies, but realizing that the hardware we possess, the brain itself, cannot change in such short periods of time (in the context of evolution) may be important in achieving a better understanding of our own decision processes as well as those of others.

The genealogy of decision processes is an important area for further research. Recent research shows that our genes predict who is likely to go to the voting booth and who may not. One cannot predict the candidate they may vote for from the genes, but the likelihood of the act of voting can be predicted. This indicates a possible hardwiring in our brains regarding decisions and the act of decision making. Similarly, people from different locations and cultures may reach differing decisions when presented with the same data. Although the effect of training and experience (human software) cannot be removed in such experiments, the notion of the hardware design of the brain as an important aspect of decision process cannot be rejected.

Now, let us think about decisions systematically and create a nomenclature that I consistently follow through the book. A decision is a conclusion reached or action taken regarding a present or future event based on past, present, or forecasted information. This definition has the following implications:

1. There are no decisions in the past. Decisions already taken are not decisions we consider in this context. As defined in finance, decisions in the past are like sunk costs. They may have an impact on the decision at hand only through the information they generated in the past.

2. Decisions may use past and present information that is observed. Decisions may also use forecasted information as necessary.

3. Decisions may or may not result in an action. In all cases, decisions relate to a conclusion reached from the information used.

4. There are no qualifications on the type of information used in reaching decisions. Information can be imprecise, historical or forecasted, based on theory or empiricism, and based on data or opinions.

Let us look at some examples.

Consider the decision by an individual investor to buy a stock in the market. The decision here is to buy or not to buy; this is a *binary decision*. It has two defined and opposite actions. This decision is not directly affected by a past decision to buy or not to buy the same stock or other stocks. This decision is taken today, and past decisions are not relevant. However, the information created from a past decision can influence the present decision. One example of this will be the risk aversion of the investor. If the investor lost money buying the same stock in the past, it may bias the current decision toward not to buy. The more relevant information for the current decision will be the financials of the company as well as future forecasts of the prospects of the company, its industry, and the overall economy. If you are a strict believer of the efficient market hypothesis (which proposes that the current price already reflects all known information), one could argue that the financials of the company do not matter as the current price already reflects such information. In this context, forecasted information requires further consideration. To forecast information, one has to follow a process. In most of the discussions here, I follow a normative process—one that is based on market expectations.

Now consider a similar binary decision, a decision to get married or not to get married. The information considered here is complex, and most of the information may be related to the other person in the marriage scenario. In certain cultures, it can get more complex as the scope of information may need to include families, clans, and in some cases entire societies. In any case, the basic elements remain the same. This decision has characteristics of an option; it gives the holder the right but not the obligation to act, assuming that the person considered for marriage is willing since the counterparty also has an option to reject the proposal. Since this action cannot be immediately remedied (such as selling the stock later after realizing it was issued by a bad company), the option to delay the decision is more important in this case. If the information considered in reaching the decision is highly imprecise (or shows a lot of variability), it may be intuitive that the right to delay the action is more valuable. The higher the variability in the information used to make the decision, the higher the chance that the decision will be delayed. There are costs of "exercising" this option as well as the present value of expected future profits from holding the asset (here the relationship with the spouse). In this case, the future profits may include the value of utility gained from companionship, availability of future options such as the ability to have children, as well as the incremental economic value gained by living together, which may include scale advantages in the purchase, production, and consumption of food, housing, and transportation as well as tax advantages such as filing a joint tax return. The cost may include lack of freedom, and if the uncertainty in benefits and/or costs is high, the exercise of this option will be delayed. Both parties may continue the relationship longer in an attempt to gather more information (and thus reduce uncertainty). If the marriage option is perpetual (does not have an expiry), it may never be exercised. However, the longer one waits, the lower the benefits. As both parties age, some of the expected benefits of the marriage may no longer exist; thus, there is a trade-off between the delay in exercise and the loss in benefits, somewhat like a call option on a stock that pays large dividends (and thus loses value).

To take this one more step, briefly consider the decision to commit suicide. This is an ultimate "put option" with no chance of reversing the effects. Since it is a "perpetual put option"—perpetual until the natural death of the person—such an option will only be exercised if the variability in future forecasts is zero. Alternately, if the present value of the future expectations of the individual is zero or negative, exercise of the put option may be optimal. As expected, most people delay the exercise of this put option as long as possible.

In 1973, Fischer Black and Myron Scholes, then at the University of Chicago, discovered a closed-form solution to price financial stock options. A financial call option (called a derivative) on a stock gives its owner the right (but not an obligation) to buy one share of the stock (the underlying) at a prescribed future time (expiry) for a preset price (strike price). The holder of the option will exercise the option at expiry only if the stock price at that time is higher than the strike price. If the stock price is lower than the strike price, the holder will simply walk away and let the option expire "out of money." The now-famous Black-Scholes equation utilizes the geometric Brownian motion (GBM), also known as random walk in physics, to model the price process of the underlying stock and market equilibrium conditions in deriving the value of an option on the stock. The stochastic processes, including the GBM, are investigated in Chapter 5, and how the fundamental arguments in options theory are derived is shown. It is, however, important to note that the closed-form solutions, such as the Black-Scholes equation, are valid only under certain conditions, such as a single option on a stock, the price of which follows the GBM stochastic process. In most decisions, we have many options that interact with each other, and it is impossible to derive an elegant closed-form solution.

Often, the stochastic processes that drive the prices of the underlying assets may not follow the highly mathematically tractable GBM. A good example of this is the price of commodities such as oil. The price of oil is affected by both demand and supply, which are both dependent on the price itself. As the price of oil goes up, demand declines (aided by conservation and higher efficiency), and supply increases (as more production capacity comes on line enticed by the higher price). These tend to drive the prices down to a long-run equilibrium level. If the price of oil goes down, the opposite happens, driving prices back up. This type of a price process is called the *mean reverting process* (MRV), and an option based on this type of asset cannot be solved by the original Black-Scholes equation.

There are two types of simple options: American and European. The American type can be exercised any time before expiry, and the European type can only be exercised on the day of expiry. Options present in decisions can be both types, and it is generally not optimal to exercise an option before expiry. One category of options typically exercised earlier than expiry is employee stock options (ESOs) given by companies to key employees as an incentive for performance. The idea is that ESOs are typically issued with a strike price equal to the current price of the stock and with an expiry date and vesting date in the future. The employee is free to exercise the option after it vests and before it expires. If employees have higher productivity and make better decisions, the stock price of the company goes up, and the employees profit from the exercise of the ESOs given to them by the company. Since ESOs behave just like exchange-traded options, it is not optimal to exercise them before expiry if one can sell them. Since employees, typically, cannot sell ESOs, they may be forced to suboptimally exercise them prior to expiry. In many cases, a large percentage of the employees' wealth may be tied up in these ESOs, and they may be forced to exercise them either to diversify their portfolios or to meet major expenses, either planned or accidental. The Financial Accounting Standards Board (FASB) recently made an accounting change that requires companies to expense these options in their financial statements. ESO valuation is described in detail in Chapter 8, taking into account past employee behavior, such as early exercise, as a function of stock price or time after award.

For the past decade, a debate has been raging among finance practitioners involved in investment decision making. A few have been advocating something called real options as an alternative to the more conventional and well-known discounted cash flow (DCF) techniques, both for valuing investment opportunities and for choosing among such opportunities when faced with budget constraints. The term *real options* was coined in the late 1980s to describe decisions on real assets that show option-like characteristics. Since these are options on real assets (as opposed to financial assets), the term seemed appropriate. As shown in the many cases discussed in this book, real options represent a way of thinking about corporate strategy, and if appropriate, it is a technique for quantifying the value of corporate assets and strategies. Stylized and academic applications of real options, however, are not sufficient to make faster and better decisions in companies. Real options have to be removed from the shackles of academics and brought to the enterprise in an understandable format that is able to solve real problems.

In the early 1990s, Hewlett-Packard Company (HP) invented and implemented a concept called "postponement." The idea behind postponement is that it is better not to fully customize a product until the last moment, that is, until the consumer is ready to pick it up and pay for it. The managers responsible for the decision did not call the method real options, but they clearly understood the value of postponing certain aspects of the manufacturing process, such as product customization, and thus keep the options alive. They decided to delay the decision to customize. The product in this case was an inkjet printer in a box, including manuals, power cords, and all the other accessories. Once a printer had been put in a Japanese box with a Japanese manual and a 100-V power supply, it could be sold only to customers in Japan; at that point HP had lost the value of any postponement options as the company has already exercised those options.

The alternative approach, which HP began to put in place in the early 1990s, was to standardize parts and assemble them in stages, reaching higher and higher levels of customization only as they approached the eventual sale. Such progressive customization had the benefit of keeping the postponement option alive (until it was optimal to exercise) while increasing standardization upstream, which provided scale advantages in manufacturing. This simple idea of postponement also led to a complete rethinking of HP's supply chains, including the location of manufacturing and assembly plants as well as warehouses.

Another example of real options is a software company that was developing software to help customize products in the computer hardware industry. By making the software specific to computer hardware companies, the software company could develop the product faster, provide "hard-wired" custom features, and go to market sooner with a product tailor-made for the targeted segment. Alternatively, it could develop more generic software as a "platform" product that could be used across a variety of industries—from furniture manufacturers to airlines. This platform product could then be customized as appropriate to the needs of the ultimate customer. After considering the choices, the company settled on investing in the platform technology and preserving the "option" to move into a variety of industries in the future. This decision, although somewhat delaying the market entry of the new software, significantly broadened its market potential.

A third example of real options at work can be seen at a large pharmaceutical company that was considering acquiring a prototype molecule from a biotechnology company. The molecule was in the very early stages of research and development (R&D), and the question was how much it was worth. Since these transactions are quite common in the biotechnology industry, one would assume that similar questions had been asked and answered numerous times in the past, and that formal valuation techniques had been developed as a result. This was not the case. Of course, companies engaging in these types of transactions have developed some "rules of thumb" based on trial and error and past experience. They are something like this:

1. Pay as little as possible.
2. Try to avoid royalties and give cash up front, if possible.
3. See if there are comparable transactions in the past and, if so, take that price and reduce it by 25% as an opening bid.
4. Do not pay more than a 25% premium over the prior price (if a prior price can be established).

The problem with such rules, however, is their failure to provide any clear guidance regarding what a reasonable price might be, especially when no precedents can be established. It is also possible that a particular rule of thumb (such as avoiding royalties) resulted from a single bad experience in the past and is not a reliable guide for future decisions. A DCF analysis of the costs, revenues, and expected in-licensing terms (at the pharmaceutical company's weighted average cost of capital) indicated that the molecule had a negative NPV (net present value). But, this result contradicted the intuition of the decision makers.

How could a product that showed such promise and had no legal issues have a negative value? Baffled by this result, they looked for answers but could not establish a normative process fast enough to help the transaction. So, they fell back on what is typically done in the industry: proxy pricing. This means that the deal terms are based on past transactions either within the company or with competitors. This is hard to do as new inventions in this industry are unique and may not have allowed usable proxies in the past.

Decisions with embedded options as described can get quite complex. For example, consider the R&D program in a pharmaceutical company. It is a long and daunting process, taking over a decade for an idea eventually to make it to a marketed drug after hundreds of millions of dollars in expenses. Typically, such programs are conducted in stages, often called phases. The duration, complexity, and required investments increase as the R&D process progresses. In each phase (or stage), a variety of decisions has to be made related to the manufacturing of the new chemical entity (the prototype chemical tried as a new drug), testing on animals and humans, and eventually filing for an approval from the Food and Drug Administration (FDA). In each phase, such decisions will result in an investment outlay. Actions taken in accordance with the decisions also result in new information. As new information arrives, we may want to change or rethink our strategy, and future decisions will be clearly be affected by the information revealed in past actions. For example, if we are unable to manufacture the new chemical in an economically viable fashion (companies may have a cost of goods sold [CGS] target for manufacturing based on the drug's pricing power and overall market potential), we may decide to abandon it even if it shows promise in some of the animal experiments. Similarly, the experiments in animal models and occasionally in human models may have shown unacceptable toxic effects that may preclude the approval and marketing of the drug, and this may force abandonment of the program. There are also other types of less drastic changes that may need to be incorporated into the program in light of new information, such as redoing or redesigning experiments, delaying certain future experiments, and if we are lucky, accelerating the program to market. It is important to note that almost always the information we have is imprecise. So, often it is not precise data gained but rather a probability distribution of possible future events. In some cases, experiments are designed to narrow the uncertainty—tighten the distribution of probabilities. At other times, experiments may be conducted to tease out information that allows the drug to be designed to increase its scope and thus reduce uncertainty.

It may not be intuitively clear why someone will design a prototype to increase uncertainty in future outcomes. Note that the R&D process is a series of interacting options, and until the drug reaches the market, the company has future options to exercise. Uncertainty is valuable for options. The more uncertain the outcome, the more valuable options are. This may be counterintuitive to those who have gotten accustomed to choosing investments to always reduce uncertainty. This may be a good way to manage if there are no future investment decisions (with options characteristics) related to the present one. But, in the presence of future decision flexibility, uncertainty can be good.

The following common themes are investigated in detail in other chapters:

1. R&D programs are typically conducted in stages, each stage designed to reveal new information that may help managers make better decisions in the succeeding stages.

2. Arrival of new information may result in changes in the R&D plan, some drastic (e.g., abandonment of the entire program) and some less drastic, such as a delay.

3. Most information gained is not precise. Information may be in the shape of a probability distribution of future events.

4. Not all experiments are designed to decrease uncertainty, and sometimes reducing uncertainty is not necessarily optimal. We want to get information that reduces intrinsic uncertainty, but we may also buy information to allow us to design a prototype to increase uncertainty in future outcomes.

I now expand the nomenclature with the introduction of some new items. I started with the definition of *decisions* in this framework and then introduced *options* as part of that framework. The staged R&D process discussed is a complex decision options problem with many different and interacting options and uncertainties. Uncertainty (risk) is unbundled into *private risks* and *market risks*. For example, the price of a traded asset such as a stock of a technology company is subject to market risks (systematic risks) that are related to the broad economy as well as the company's industry. The process also has private risks, such as the risk of losing a specific chip design patent the company holds in a lawsuit. The private risks (unique risks or unsystematic risks) can be diversified away by holding a large number of companies in a portfolio. Intuitively, we can understand that the private events to which any specific company is exposed are not correlated to other companies; thus if there are many different companies in a portfolio, the shock of a private event can be diluted and, in the extreme, completely diversified (as there may be positive and negative shocks). As such, the private risks will not be priced by the market, that is no one will pay a premium to assume private risks as they can be costlessly diversified in a well-functioning and broad stock market. The market risk (systematic risk), however, is correlated with the overall economy and thus cannot be diversified by having a large number of differing companies in the portfolio. For example, recession in the U.S. economy affects (in differing degrees) most companies in the economy and thus is a systematic or market risk.

In the decision options framework, private risks and market risks are treated differently just as in traditional frameworks such as the single-factor capital asset pricing model (CAPM). Even though most companies practice a version of CAPM in discounting cash flows to determine the NPV of investments, it is generally applied in an *ad hoc* fashion without a clear delineation of private and market risks. In CAPM, systematic risks are captured in the discount rates and private risks in probability adjustments to the cash flows in the numerator. Often, in the practice of DCF analysis both private and markets risks are bundled into one with arbitrary adjustment to the discount rate to account for both together. In many cases, an internal rate of return (IRR) is calculated from the cash flows and then compared against a threshold level. An IRR generally incorporates private and market risks into a single number and is thus inconsistent with traditional theories such as the CAPM.

This can be made more personal by considering a real-life situation of a decision options problem with private risks embedded. Suppose you would like to buy a piece of land and construct an office building to rent to others. The decision to be made is how much to pay for the land today. Perhaps two options can be seen: the option to buy the land and then an option to construct the building. In the absence of any additional data, this decision options problem can be solved as two sequential and interacting options driven by market processes such as real estate prices and rental rates in the area. But, it is made more interesting by introducing a private risk. An example of such a risk would be the probability of a zoning change. The land is currently zoned commercial, and there is a chance, say in 2 years, that the zoning will be changed to residential. Also assume that the land acquisition decision must be made in 1 year (the first option) and the construction decision in 3 years (second option). Between these two options, there is now a private risk—the risk of a zoning change—that is not associated with market processes such as real estate prices and rental rates. The chance of a zoning change is really at the whim of the county regulators, who are notoriously fickle and make decisions uncorrelated with the broad market. This decision option problem has two options connected by a private risk. For those who seek pain, additional private risks can be imagined, such as the risk of unearthing hazardous materials while excavating the land for the foundation of the office building. Such a risk is also private and is unrelated to the market.

Some components of decisions can be represented as swaps or options to swap. For example, consider an electricity generation plant. First, consider a base load plant, a plant that operates continuously to provide a base level of power to the grid. The demand for power will always be greater than what is provided by the base load plants, and they have to operate without stoppage. The plant is fired with oil and produces electricity that is supplied to the grid. If the price of electricity is greater than the cost of fuel and maintenance, the plant operator makes a profit. If not, the plant will run at a loss. Such a plant can be represented as a swap between electricity and fuel. If profits and losses every hour are examined, the transaction will be an hourly swap during the time of operation (until the next expected outage). If the price of electricity is higher than the cost of operation (including fuel and personnel costs), the plant will make money. Otherwise, it will lose money.

We can also think of a peaking plant: a plant that produces electricity when demand is higher than what is supplied by the base load plants. Let's make this interesting by introducing a dual-fuel plant; it can take both natural gas and oil as fuel. Also assume that this plant was designed to be flexible; this plant can be turned on or off at will. Now there is a very interesting decision options problem. The plant operator can chose from gas or oil to fire the plant (the operator may use the cheaper fuel) and decide to operate or shut down, depending on the price of electricity. Electricity is a unique commodity that does not yet allow a very efficient storage mechanism. This means that electricity needs to be used once produced or is only produced when there is a demand. This makes the electric grid a dynamic place where the price can wildly fluctuate. One does not want a power failure in the middle of a heart transplant operation, so if the demand is there and the supply is limited, price will climb. This is a custom-tailored situation for the operator of the dual-fuel flexible plant. Every hour the operator considers prevailing prices and decides to start the plant or shut down. If plant will be operated, a decision on which fuel to use also needs to be made. In many cases, there may be contracts and fuel hedges in place that maximize the overall profits of the company (that owns the plant). This may make the plant operating decision a very complex one. In any case, the plant can be visualized as a series of options to swap between electricity prices and the lowest-cost fuel prices. The decision options framework allows a variety of such derivative instruments to interact in a complex decision process.

Now you are ready to explore the decision options framework—a framework that allows better decisions to be reached systematically, to represent all components of a decision (e.g., options and risks) in a standardized fashion, and to utilize a dynamic process to incorporate new information as it becomes available. Always remember the two important ingredients of a decision options problem: uncertainty in inputs and decision flexibility in present and future decisions. When uncertainty and flexibility are present, decisions will have options like characteristics, and traditional frameworks that ignore one or both fail to provide reasonable answers. The utility of a generalized framework that always considers uncertainty and flexibility will become obvious as we take this journey together.

3

Decisions as Predetermined Outcomes, Options, and Risks

This chapter dissects decisions into three major building blocks: predetermined outcomes, options, and risks. These three components form a decision options framework. We are only concerned about the future, and any decisions or occurrences in the past have no impact on future decisions. In the traditional financial jargon, the past may be considered as "sunk" and thus irrelevant for future decisions. Each of these components has distinctly different characteristics.

A *predetermined outcome* is something that is expected with certainty in the future. As many know, death and taxes fit this criterion well. Although there are no future events with a probability of 1.0, in the decision options framework these are outcomes that are "contracted" to happen. For example, if we enter into a contract that we pay somebody $1 now and the person pays us back $1.10 in 1 year, both the cash outflow of $1.00 at time 0 and the cash inflow of $1.10 at time 1 are predetermined outcomes. In the decision options framework, we further generalize predetermined outcomes and the time of occurrences as not constants but probabilistic. For example, we can enter into a contract by which we pay somebody $1 now (both time and amount are constants) and the person pays us back an amount that is a sample taken from a normal probability distribution with an average of $1.00 and standard deviation of $0.50 at a time that is a sample from a lognormal distribution with an average of 1.0 year and a standard deviation of 3 months. In this case, the cash inflow in the future and the time of inflow are both probabilistic. Chapter 4 treats probability distributions in detail; it will suffice to remember now that a probabilistic outcome can also be a predetermined outcome as long as the characteristics of the distribution are predetermined. In this complex contract, we predetermined that the cash inflow will occur at a time that is a sample from a lognormal function with predetermined characteristics (average and standard deviation), and the amount of inflow is also predetermined as a sample from a normal distribution with predetermined characteristics (average and standard deviation). We assume that all contracts will be honored, and we live in a society with well established contract laws and property rights.

An *option* is a right but not an obligation to take certain action in the future. The action can be taken at a predetermined time (European option) or anytime before a predetermined time (American option). For example, a contract to pay $0.10 now for the right to claim $2.00 in 1 year by paying €1 is an option that gives us the right to buy $2.00 for €1 in that timeframe. This option will not be exercised if the dollar continues its depreciation and the price of €1 is more than $2.00 in 1 year. If the price of €1 is less than $2.00, we will exercise the option and pocket the difference. This is a simple and self-standing option with no interactions and can be solved with available closed-form solutions.

The third component of the decision options framework is *risk*. Risk can come in different flavors. The most basic type of risk is a probability that an event will happen. For example, we can say there is a 50% chance that it will rain tomorrow where you live. So, the risk of rain happening (or not happening) is 50%. This is a binary event with a constant probability of occurring. In the traditional finance based on the capital asset pricing model (CAPM), we identify two distinct types of risks: systematic (market) and unique (private). The term *market* is used throughout this book to mean the universe of all investable securities, and it has the same meaning as the market in traditional finance, the correlation against which determines the β in the single-factor model of CAPM. It is sufficient to note that we are used to assuming the Standard and Poor's (S&P) 500 or the Wilshire 5000 index of traded stocks in the United States as a fair representation of the market. However, as the percentage of the World gross domestic product (GDP) created by the United States declines (in both nominal and real terms as the dollar loses value), it will increasingly become a less-accurate proxy for the "market." In any case, the systematic risk cannot be diversified away, and the holders of this type of risk will expect a reward commensurate with the magnitude of the risk taken. The other type of risk is called a *private risk*. If the event is uncorrelated with the market (such as the probability of rain tomorrow), it can be fully diversified away by combining a large number of uncorrelated events. Since the outcomes of these uncorrelated events are unrelated to one another, the effect of one may be cancelled out by another, and thus the "portfolio" of a large number of such events will not have any private risk. For example, consider three uncorrelated events: the chance of rain, the chance of finding a parking spot close to the railway station, and the chance of finding a friend with an umbrella. If you set out to catch the train without an umbrella, the net effect of your getting drenched by the rain depends on the probability of all of the uncorrelated events. If it did not rain, you reach the train without any problem. If it rained but you secured a parking spot close to the station, you avoid an encounter with rain. If it rained and there were no parking spots close to the station but you found a friend with an umbrella, you accomplish the mission. With the availability of these three uncorrelated events, the risk of being drenched by the rain is substantially reduced. This portfolio effect of the ability to diversify uncorrelated risk is at the heart of the traditional financial theory.

In the decision options framework, private risks are incorporated in different ways. Private risks can lead to a probabilistic outcome as discussed. They can also be incorporated into a time series as a sudden change (a jump). In

Chapter 5, stochastic functions (time series) such as random walk are detailed. A time series shows the value of a variable (such as the Dow Jones Industrial Average) over time. Readers may be familiar with time series charts shown in various business newspapers and on television. They show how a price has changed over time. It is easy to draw prices from the past (as we have already observed the prices), but forecasting them into the future is a different story. Although many pundits claim that they have perfect foresight predicting prices of stocks and commodities, the evidence for this is scant.

We can, however, create a price process that can model the future progression of prices. Such a process does not tell us the precise outcome in the future but can incorporate the characteristics of the expected evolution of prices in the future. One such process is the random walk. A random walk is expected to follow certain mathematical characteristics and does not include a jump. This is because most often prices evolve smoothly. However, a private risk (which is uncorrelated with the broad market) can be introduced into such a smooth process as a sudden deviation from the expected next value (a jump). For example, the Dow Jones Industrial Average may follow a random walk but can also show jumps (extreme events) such as the 9/11 disasters.

As mentioned, the decision options framework allows combining options, predetermined outcomes, and risks in making a decision. The framework is generalized to the highest level so that any combination of these components that may interact with each other can be solved in a single framework. Although interactions between predetermined outcomes and risks are easily incorporated with simple mathematics, interactions among options are a more complex phenomenon. Traditional finance based on the CAPM deals only with predetermined outcomes and risks. For example, when we calculate the NPV (net present value), we take cash flows (which are by definition predetermined) and adjust them for noncorrelated risks (private risks). Typically, we multiply a cash flow by the probability of that cash flow occurring and then discount the probability-adjusted cash flow by a rate that reflects the systematic risk of the cash flow discounted.

For example, consider an investment proposal that requires an investment of $100 now and expects a payback of $200 a year from now. The $200 that we expect back carries a systematic risk similar to the market (a β of 1.0). The β is a measure of systematic risk. The β (as shown in the next equation) is a function of the standard deviation of returns of the security σ and that of the market σ_m as well as the correlation of the return of the security with the market ρ:

$$\beta = (\sigma / \sigma_m)\rho$$

The β can also be expressed as the ratio of the covariance of the asset's return against the market and the variance of market returns as follows:

$$\beta_{am} = Cov(R_a, R_m) / Var(R_m)$$

Note that β depends on both how the securities returns are correlated with the market and the volatility (standard deviation) of the security's returns. A security perfectly correlated with the market (correlation of 1.0) but showing volatility twice that of the market will have a β of 2.0. Similarly, a security that is showing volatility in returns similar to the market but a correlation of 50% will have a β of 0.5.

Assume that the return expectation on a market portfolio is 10%. This means that if we buy all securities in the exact proportion as in the "market," we expect to make a return of 10%. If we can find a single security that is perfectly correlated with the market and shows variance similar to the market, such a security will have a β of 1.0 and thus an expectation of return similar to the market (10%). Further assume that there is a binary event that may happen 6 months from now. This event either maintains the cash inflow expectations or drops that to zero catastrophically. Note that $200 is an *expectation*—an average of all possible payoffs that can occur in various future states of the market. In this case, we can express net present value as

$$NPV = -100 + p \times 200 / (1+r)$$

where p = .5 and r = 10%

All of the parameters in the NPV equation are predetermined. In the preceding NPV equation, the cash flows are presupposed to occur with certain probability. We accept the cash flows as they come; we do not have any flexibility to modify them based on a future state (no optionality). This does not mean that the cash flows are definite to occur or do not carry any risk. In fact, the $200 payback is not at all ensured. It carries two types of risks: the risk of a catastrophic failure resulting in a $0 payback (regardless of the state of the market) and the risk of an outcome of more or less than $200 as a function of the state of the market. However, we assign specific factors representing these risks today (we adjusted the $200 expectation by 50% to account for the binary event and then discounted the probability-adjusted cash flow by 10% to account for the market risk) and calculated an NPV assuming that we have no flexibility to change them in the future.

Options are different. Options allow us to pick a more favorable cash flow in the future depending on a future state. For example, consider a modified investment proposal that requires a payment of only $10 today for the right to make a decision to buy the payback 1 year from now at a cost of $200. The probability of the binary event is maintained as before. As before, we can write the following NPV equation:

$$NPV = -10 + p \times OptionValue$$

where p = .5

We need to now calculate the *option value*, the value of the option today to buy the future payoff of the project at a cost of $200. In options parlance, $200 is the "strike price". We have 1 year to "exercise" this option. What we get ("asset") is a cash flow that is expected to be $200. In the real world, the project payoff (asset) can take on a variety of values based on the state of the economy. For the traditional NPV calculation, the possibility of a variety of values for the project payoff was not relevant as we were only interested in the expectation (or average payoff). The shape or the distribution of the payoff was also of no consequence; the average was sufficient to calculate the NPV. However, in the case of the current example that includes an option, the average is not enough as we have the right (but not an obligation) to exercise the option in the future after having observed the state of the project (and the market). This gives us the flexibility to alter the cash flow in our favor.

Options pricing is discussed in more detail in the succeeding chapters. It is sufficient to remember now that when options are present, the distribution of the future payoff figures in the valuation as well. To make matters more interesting, if multiple options are present in the same decision options problem, they interact with each other, and such a problem cannot be solved with closed-form solutions such as the famous Black-Scholes formula for options pricing. It is also worth noting that closed-form solutions for options pricing have constraining assumptions such as the stochastic characteristics of the underlying cash flow (the project in our example) and are not universally applicable. In the decision options framework, we generalize all factors that affect the price of the option as well as the factors that affect predetermined outcomes and risks.

To summarize, the decision options framework is a generalized framework that allows the solution of any combination of options, predetermined outcomes, and risks. All three of these constructs have meaning that is much more general than what is traditionally meant by them. For example a predetermined outcome just means that the characteristics of the outcome are predetermined, and we have no optionality (flexibility) to alter them in the future. Risks encompass all types—correlated and uncorrelated to market. Options include future decisions that have flexibility, and they can be based on evolving parameters that show uncertainty of many different kinds.

4

Private Risks and Probabilities

A *private risk* is uncorrelated with the market. The risk of rain tomorrow and the risk of an accident at 5:00 p.m. today on Interstate 95 are private risks (unique risks). Equally unique is the risk of the failure of a dam or the risk of obtaining unfavorable data in a clinical experiment on a particular drug of a pharmaceutical company. These types of risks are not related to what is happening in the general economy. All the risks mentioned will have the same chance of happening whether the economy is in recession or growing rapidly.

The decision to carry an umbrella clearly depends on the risk of rain. We can incorporate this risk into our decision options framework as a binary outcome. A binary outcome has two possible states (it rains or it does not rain); one has a probability p and the other a probability of $1 - p$ of occurring. For example, we can look at how many days in the last 30 days it rained. If it rained 15 of the 30 days, the risk of rain is 50%.

Risks can also be on variables that can take many different values (not just binary). For example, we can plot the maximum temperature on all days in May in Manhattan for the last several years. We can create a probability distribution (or a histogram) of maximum temperatures.

Figure 4.1 shows one possible scenario. On the x axis, we have the temperatures (°F); on the y axis we have the probability of that temperature occurring on any day in May. It shows an average temperature of 62°F and a temperature range of 45 to 90°F. This is a probability distribution of temperature. It also shows that it has a standard deviation of 9. We also calculated what is called higher moments of this distribution, such as skewness and kurtosis. The distribution shows a skewness close to 0 and kurtosis of over 5. Students of probability know that a normal distribution has a skewness of 0 and kurtosis of 3.

So, this temperature distribution shows skewness close to normal but a higher kurtosis (tendency to have thicker tails; higher probability of extreme events than what is expected if the temperature was normally distributed). A positive skewness (higher than 0) would have implied a tendency to have significantly higher than average temperatures than normal and a negative skewness (less than 0) the opposite. We can also draw out the cumulative probability distribution of the same temperature observations (Figure 4.2).

The x axis of Figure 4.2 shows the various outcomes of temperatures, and the y axis shows the cumulative probability of that temperature. Cumulative probability as represented here is the probability that the temperature on

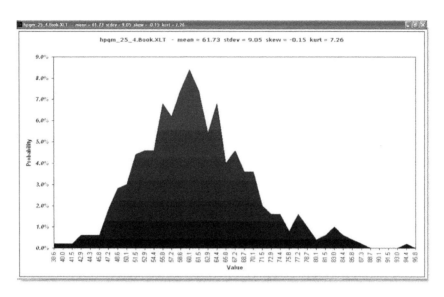

FIGURE 4.1
Temperature histogram.

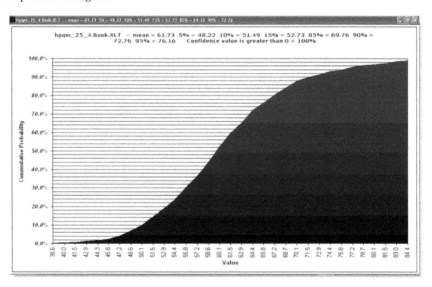

FIGURE 4.2
Cumulative probability distribution of temperature.

any specific day is lower than a specific number. For example, the cumulative probability for a temperature of 70°F is 85%. This means that there is a 15% chance that the temperature on any specific day in May will be higher than 70°F. We can also create confidence intervals. For example, we are 90% confident that the temperature will be between 52 and 73°F.

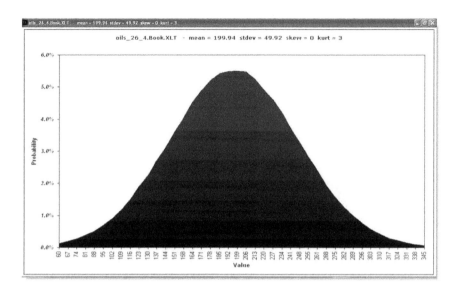

FIGURE 4.3
Normal distribution with mean of 200 and standard deviation of 50.

The examples in this book deal with three primary types of probability distributions: normal, lognormal, and triangular. A normal distribution is the classical "bell curve" in statistics. This is sometimes also called *Gaussian distribution* and is totally symmetrical about its mean (average). Given in Figure 4.3 is a normal distribution function with average of 200 and standard deviation of 50. It has a skewness of 0 and kurtosis of 3.

We use the normal distribution when a large number of observations show a roughly symmetrical shape around an average. A normal probability plot can be used to assess the normality of a data set. The equation that represents the probability of an outcome in a normal distribution (also called the probability density function) can be represented as

$$P(x) = \frac{1}{\sqrt{2\Pi}} e^{\frac{-(x-\mu)^2}{2\sigma^2}}$$

where μ = average, and σ = standard deviation.

A predetermined outcome may be a sample from a normal distribution. If we enter into a contract to pay \$100 today and receive a sample from a normal distribution of average 200 and standard deviation 50 at 1 year from now, the value of the contract will be

$$NPV = -100 + N(200, 50)/(1+r)$$

where $N(200,50)$ represents a sample from a normal distribution of average 200 and standard deviation 50.

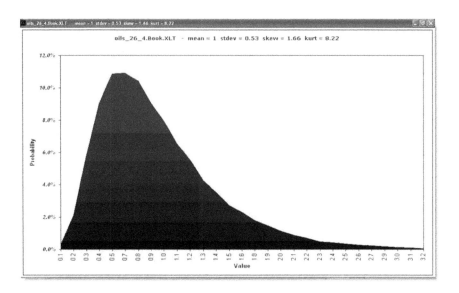

FIGURE 4.4
Lognormal distribution with mean of 1.0 and standard deviation of 0.5.

The discount rate *r* in this equation will be the risk-free rate as the project's cash flow 1 year from now is uncorrelated with the market and hence carries no systematic risk. Normal distributions may be used to represent items such as the probability of the total number of experiments needed to reach a milestone or the probability of finding a certain level of impurity in a chemical sample when a large number of previous readings showed a probability distribution that is roughly normal.

The second type of distribution that is typical in business problems is the lognormal function. The lognormal distribution is a single-tailed distribution of a random variable whose logarithm (to the base e) is normally distributed. In other words, if variable *X* is lognormally distributed, then ln(*X*) is normally distributed where ln represents the logarithm to the base e (natural logarithm). Figure 4.4 shows an example of a lognormal distribution with mean (average) of 1 and standard deviation of 0.5. It clearly shows a positive skew (in this case 1.7) as well as a kurtosis (thicker tail) above 8.0 (which is higher than the expected 3.0 for normal function).

The probability density function of a standard lognormal function can be represented as

$$P(x) = \frac{e^{-((\ln x - \mu)^2 / 2\sigma^2)}}{x\sigma\sqrt{2\Pi}}$$

$$Average = e^{(\mu + \sigma^2 / 2)}$$

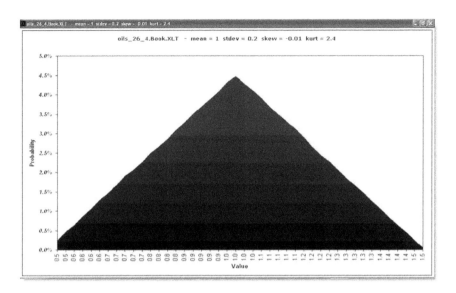

FIGURE 4.5
Triangular distribution with mean of 1.0, minimum of 0.5, and maximum of 1.5.

Lognormal functions may be used to represent the time taken to complete an activity or the total cost to completion of an activity. If we enter into a contract to pay $100 today and receive a cash inflow of a sample from a normal function with average 200 and standard deviation 50 at a time that is a sample from a lognormal function with average of 1 and standard deviation of 0.5, the net present value (NPV) of this transaction can be represented as

$$NPV = -100 + N(200, 50)/(1+r)^{\log norm(1,0.5)}$$

Since this cash flow is not correlated with the market, the discount rate r will be the risk-free rate (as β is zero).

A third type of probability distribution that may often fit historical data well is the triangular function. Although this is not as mathematically tractable as the previous two functions, it is simpler to visualize and understand. The probability density function of a triangular distribution looks like a triangle, hence its name. For example, the probability density function given in Figure 4.5 is for a distribution that has an average value of 1, minimum value of 0.5, and maximum value of 1.5.

Let us further complicate the transaction described in which we elected to receive a sample from the normal distribution of $N(200,50)$ at a time represented by a sample from the lognormal distribution, lognormal(1,0.5). Assume that the amount we receive is multiplied by a sample taken from a triangular distribution with an average of 1, maximum of 1.5, and minimum of 0.5. In effect, we are saying that the project payment may be reduced as

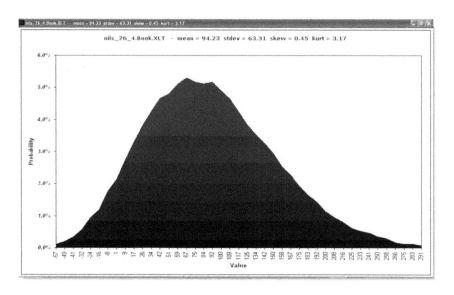

FIGURE 4.6
NPV distribution of the contract.

much as half, but we can also get a bonus as much as half of the project's expected value based on the outcome of an external event. We can imagine this as a bonus structure in which we take the actual project cash flow and increase it (positive bonus) or decrease it (negative bonus). The NPV of this contract can be represented as

$$NPV = -100 + tria(1, 1.5, 0.5) \times N(200, 50) / (1 + r)^{\log norm(1, 0.5)}$$

For a risk-free rate of 3.0%, the NPV distribution of this contract is as given in Figure 4.6 with an expected value of $95.

In this contract we deal with a private risk—the risk of a bonus (positive or negative). We also have two predetermined outcomes: the timing and the amount of the project cash inflow. We also do not have any optionality (flexibility) as the entire problem is composed of predetermined outcomes and private risks.

It is worth mentioning that those who are used to applying the capital asset pricing model (CAPM) for financial analysis may be uncomfortable using the risk-free rate for discounting any cash flow. Remember that a strict application of the CAPM will require a discount rate based on the correlation of the asset in question against the broad market. If no correlation exists, the correct discount rate is the risk-free rate and not the weighted average cost of capital (WACC) of the company, which represents the average risk of all projects in the company. It is also worth mentioning that the WACC is not a suitable discount rate for a single project in the company unless that project has the average risk of all projects in the company's portfolio. The CAPM requires a discount rate commensurate with the risk of the project considered and not the risk of the portfolio (the company) in which the project resides. A large percentage of the discounted cash flow analyses conducted in corporate finance may use the WACC or a corporate prescribed rate (such as 10% or hurdle rate) in discounting cash flows from all projects. This is for convenience only. NPV created from discounting project cash flows at WACC will overvalue projects with a higher systematic risk (which requires a higher discount rate) and undervalue projects with a lower systematic risk (which requires a lower discount rate). Often, the cash flows of projects are arbitrarily adjusted to account for such disparity. Both the discounting at average discount rate and the subsequent adjustments of cash flows to attribute systematic risk differences in projects take us away from the spirit of traditional finance theory, the CAPM.

5

Market Risks and Stochastic Processes

This chapter introduces stochastic processes (time series) and explains how they are related to decisions as well as risks. First, let us revisit the traditional finance theory: the capital asset pricing model (CAPM), which is a single-factor model. As discussed previously in this book, the CAPM can be used to determine the necessary rate of return on an asset. The private risks of the asset can be diversified away and will not be priced by the market, so the only relevant risk is the systematic risk that is represented by β. Mathematically,

$$E(R_a) = R_f + \beta_{am}(E(R_m) - R_f)$$

$E(R_a) =$ Expected return on asset a

$E(R_m) =$ Expected return on market

$R_f =$ Risk-free rate

$\beta_{am} = \beta$ of asset a against market

$$\beta_{am} = Cov(R_a, R_m) / Var(R_m)$$

$Cov(R_a, R_m) =$ Covariance between asset returns and market returns

$Var(R_m) =$ Variance of market returns

When we utilize the CAPM to price an asset, we need to calculate a β first. To do this, we need the covariance of the asset's return against the market's return as well as the variance of market returns. Note that the "market" is the summation of all investable assets and by definition nearly impossible to observe. So, a strict application of CAPM is impossible, but in most cases we select imperfect proxies for the market such as the Wilshire 5000 stock index in the United States.

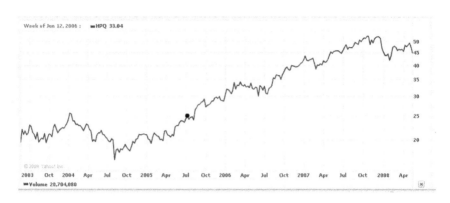

FIGURE 5.1
HPQ stock price from 2003 to 2008.

Let's look at HPQ (Hewlett-Packard Company) stock as an example. Assume that Wilshire 5000 is a good proxy for the market. By using the daily returns of HPQ and the market, we can calculate a β. Assume that the β of HPQ is 1.4. Further assume that the expected yearly return on the market is 8%, and that the risk-free rate is 3%. In this case, we can calculate the expected return on HPQ stock:

$$E(R_{HPQ}) = 3\% + 1.4(8\% - 3\%) = 10\%$$

Figure 5.1 shows the price history of HPQ for 5 years. Clearly, in some years HPQ underperformed, and some other years it overperformed compared to the expected return of 10% (based on the market benchmark and its own β).

The weak form of the efficient market hypothesis (EMH) developed in the 1960s at the University of Chicago asserts that the prices of all traded assets reflect all known information in the current price; thus, past prices are not useful in predicting future prices. This conceptually simple and elegant idea has stood the test of time. Although there are many challenges to the empirical validation of EMH, in a world with a large number of participants using widely available technologies, it is difficult to conceptually challenge the idea that all information gets reflected in the prices of traded securities reasonably quickly.

In this context, we can now investigate a stochastic process called random walk. *Random walk* is a general stochastic process (time series) in which the position of a particle (which is doing the random walk) in the next moment solely depends on its position now and some random variable. What positions the particle assumed earlier provide no additional information that is not available in its current position. A case of the random walk is called geometric Brownian notion (GBM). GBM satisfies the requirements of the EMH and has some elegant mathematical properties that make it very attractive as a model for the evolution of the price of any traded asset. The GBM stochastic process can be represented by the following differential equation:

$$dS_t = \mu S_t dt + \sigma S_t dW_t$$

In simple terms, it means that the change in the price of the asset dS_t depends on two factors. The first term, $\mu S_t dt$, is called the *drift term*. It is a function of the current price and a constant μ that is called *drift*. The second term is called the *diffusion term* and is a function of a constant σ that is called *volatility*, the current price as well as the derivative of the Weiner process (please see below). The analytic solution of the differential equation is

$$S_t = S_0 e^{(\mu-\sigma^2)t+\sigma Wt}$$

where
  S_t = Value at time t
  S_0 = Value at time 0
  μ = Drift
  σ = Volatility
  W_t = N(0,1) $\sqrt{t}$
  N(0,1) = A sample from a standard normal distribution (average = 0, standard deviation = 1) as shown in Figure 5.2. It has skewness of 0 and a kurtosis of 3.

The analytical solution of the GBM can also be expressed as follows:

$$Ln(S_t) = Ln(S_0) + (\mu - \sigma^2)t + \sigma N(0,1)\sqrt{t}$$

The random variable $Ln(S_t)/Ln(S_0)$ is normally distributed with a mean $(\mu - \sigma^2)t$ and a standard deviation $\sigma\sqrt{t}$.

Let us revisit the Hewlett-Packard example. We derived an expected return for HPQ to be 10%. If HPQ stock is following GBM, it will show a drift of 10%. Assume that the volatility of HPQ stock is 30%. With these two parameters, the drift (10%) and volatility (30%) applied on the current price of HPQ (assume 20), we can project future (possible) price paths for HPQ using the analytical formulation of the GBM.

Figure 5.3 shows 25 different stochastic simulations of HPQ price over 5 years starting from the price of 20. The only two parameters needed to conduct this simulation are the drift and volatility and the assumption that HPQ price evolves according to GBM, a type of random walk. From this point, the terms random walk and GBM are used interchangeably to mean the GBM

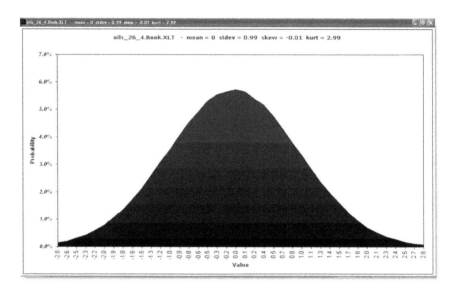

FIGURE 5.2
Normal distribution with mean of 0.0 and standard deviation of 1.0.

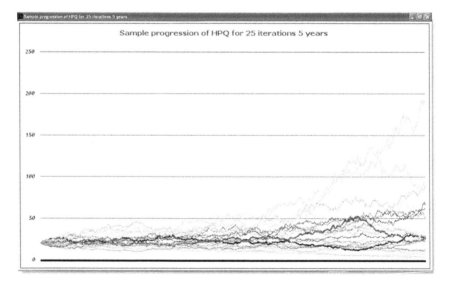

FIGURE 5.3
GBM stochastic simulation of HPQ price.

stochastic process that is used to approximate the evolution of the price processes of all traded assets.

To better understand why we use GBM as a proxy for price progression of traded assets, we have to revisit the EMH. The idea of efficient markets was first expressed by the French mathematician Louis Bachelier in his doctoral dissertation, "The Theory of Speculation." Nearly 65 years later, Eugene Fama at the University of Chicago published his thesis, revitalizing the idea. Five years later, Fama formalized EMH further and provided a review of both theory and evidence. He also provided three levels of the EMH: weak, semistrong, and strong forms.

The idea of efficient markets seems to create strong emotions in its proponents and challengers. However, at the heart of EMH is a very simple notion. In the modern world, information travels fast and reaches the participants in the market of traded assets. Not all participants may reach the same conclusion regarding the effect of such new information on a particular asset, but when there are a large number of participants (all trying to maximize utility but not necessarily completely rationally), the combined effect of actions of all participants moves the price of the traded asset to an efficient level. That is, the price reflects the information efficiently, and no participant can take advantage of new information to create arbitrage profits.

Let's look at EMH in a more practical context. I am sure most readers of this book have invested in the stock market at one time or the other. If we watch any business show on TV, we find many experts able to "precisely" predict where the markets are going tomorrow, next week, and next year. Many are even able to "pick" specific stocks and predict where the prices will be in the future. In fact, there is an entire industry of buy-side analysts providing future prices (target prices) for all kinds of securities. It is nearly impossible to find an analyst admitting that he or she does not know what the price might be in the future. Most traders are confident in their predictions and their ability to make money on those predictions. In spite of this overflowing wisdom of market gurus, evidence shows that few are able to "beat" the market indices on a consistent basis. Note that beating the market requires the calculation of a metric, α, that stands for risk-adjusted excess returns. Remember that if one bought a stock with a β of 2.0 and the market went up 10%, the stock is expected to go up 20% (assume risk-free rate is zero). Raw returns are not sufficient to assess whether the stock picker has beaten the market. The return has to be adjusted for the risk and an excess return (α) needs to be calculated. It is true that some are able to create α sometimes, but the persistence of α is very rare. This means that we are in a world in which there is not much information advantage to market participants, and the prices are generally efficient.

For completeness, also understand the three levels of EMH as proposed by Fama. The *weak form* of EMH states that current prices reflect all past information. This means that the billions of dollars expended in technical analysis in predicting stock prices is a waste. Patterns in past prices do not hide any information (that is not already in the current price) and thus are not useful at any level in predicting future prices. The *semistrong form* takes it to the next step and asserts that all current information is already reflected in the prices. This means that the billions of dollars expended in fundamental analysis of the financial statements of companies represent a waste. Prices adjust instantly to an efficient price as soon as a company publishes its financial statements; thus, any type of analyses of the financial data after they are already known to the market is not going to be useful in predicting future prices. The *strong form* of the EMH makes the analysis complete by asserting that any information (published or unpublished) is already reflected in the current prices. This means that not even insiders who may have proprietary information can take advantage of it. In other words, insider trading (which is illegal) is not really useful in creating excess profits.

Empirical evidence is strong for the weak and semistrong forms of EMH. The fact that many executives in recent times ended up in jail for insider trading (and making tangible excess profits) implies that markets are not quite strong-form efficient. However, we can comfortably assume that markets are semistrong efficient as persistence of overperformance is very rare among money managers and market forecasters who use both technical and fundamental analyses. This means that the current price reflects all the past and present information, and that only "new or unknown (insider) information" will have a predictable impact on future prices. Since we do not know the nature and timing of new information that may arrive in the future, it is impossible to predict future prices of traded securities, such as the common stock of a company. In practice, new information arrives randomly; hence, the prices will also move in a random fashion. This may not be a comfortable thought to the many technical and fundamental analysts who make a living either by "charting" or by "financial statement analysis." But, the fact that none of the experts has been able to "corner" the market proves that it is not that easy.

GBM satisfies the characteristics of price movement of a traded asset if the market for that asset is efficient and at least the weak form of EMH holds. As we have seen, GBM has two components: a drift component (constant) and a volatility component (random). We established the reason for the volatility component through the EMH. The drift term represents an "expected return" of the asset. Since there is a risk to any asset, nobody will hold the asset unless the asset is "expected" to return at a level commensurate with the systematic risk taken. As discussed, the expected return of an asset is dependent on the level of systematic risk (risk correlated with the market) in the asset.

$$\ln(S_t) = \ln(S_0) + (\mu - \sigma^2)t + \sigma N(0,1)\sqrt{t}$$

where

$(\mu - \sigma^2)t$ = Drift component that represents the expected return

$\sigma N(0,1)\sqrt{t}$ = Random component that represents random arrival of new information

$N(0,1)$ = Standard normal function

From this point, assume that any traded noncommodity asset in the public markets follows GBM. This is a practical assumption and one that is well supported by both theory and empirical evidence.

There are certain classes of assets, such as commodities (e.g., crude oil), that show mean reversion to a long-run average value. For example, in the case of oil, mean reversion happens because higher current prices of the commodity attract higher levels of investments in exploration, which results in a higher supply in the future. This is coupled with lower demand aided by conservation and diminished use. Both of these factors will dampen prices, pulling them down to a long-run mean. Similarly, lower current prices result in underinvestment and subsequent lower supplies. Simultaneously, demand will increase as use increases and efforts on conservation decrease. This will drive up prices in the future. This tendency for prices to decline after they have risen and increase after a collapse is called *mean reversion* and is seen in the price processes of commodities such as oil. This is also the case for asset classes such as currencies and interest rate futures for similar reasons. The differential form of the arithmetic mean reversion process can be represented as

$$dS_t = \eta(m - S_t)dt + \sigma dW_t$$

where
η = Rate of reversion
m = Long-run average
σ = Volatility
$W_t = N(0,1)\sqrt{t}$

For the mean reverting stochastic process given, we can calculate a *half-life*, which is the time taken to retrace half the way to the mean value once the price has deviated from it. Mathematically, half-life can be expressed in the following way, where H is the half-life, and η is the rate of mean reversion:

$$H = \ln(2)/\eta$$

Given in Figure 5.4 is the yearly price of crude oil (inflation adjusted) from 1949 to 2006. Also, Figure 5.5 shows the simulation of crude oil prices with the following parameters: $20 starting price, $20 long-run average, 20% volatility, and 5-year half-life Throughout this book, we designate this type of mean reverting process MRV.

Note that mean reversion does not mean that prices will be pulled back fast to the mean. It really depends on the reversion rate (or half-life). The higher the reversion rate, the faster the price will move back to the long-run average. In the simulation, we assumed that the half-life was 5 years, which is sufficient to show significant excursions away from the long-run mean for

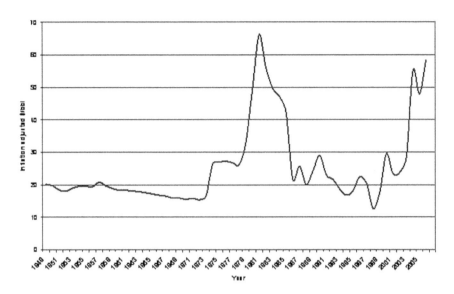

FIGURE 5.4
Inflation-adjusted crude oil prices from 1949 to 2006.

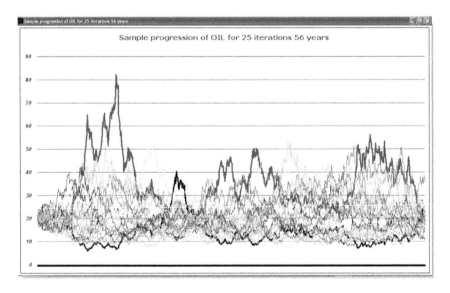

FIGURE 5.5
Stochastic simulation of crude oil prices for 56 years.

extended periods of time. If the reversion rate is low, it will be difficult to differentiate the mean reversion process from its cousin, the GBM. However, most commodities have reasonably high reversion rates, and modeling the price processes of commodities such as oil cannot be done by GBM. In addition, interest rates and currencies also show mean reversion for other reasons and typically cannot be modeled by GBM.

Interest rates also show mean reversion characteristics due to the expansion and contraction of the overall economy and business cycles. Interest rates tend to rise (more demand for money) in rapid economic expansions. This will also be reflected in the monetary policy as the U.S. Federal Reserve (FED) may move to slow a fast-accelerating economy by raising borrowing rates for the banks or removing liquidity from the markets through other open market operations. The reverse may happen in economic contractions. Similarly, currency exchange rates also show mean reversion. Intervention by central banks is one reason. A central bank of a country may act to shore up its falling currency by buying it or slow fast appreciation by selling it. In an integrated macroeconomy, a depreciating domestic currency will result in stronger exports and improving current account surplus, which in turn begin to move the currency back to an equilibrium level. So, both the interest rate and the currency exchange rates are typically modeled with mean reversion properties.

Another form of mean reverting stochastic process can be written as follows:

$$\ln(dS_t) = \eta(\ln(m) - \ln(S_t))dt + \sigma dW_t$$

In this case, it is not S but $\ln(S)$ that is following the mean reverting process. As can be seen from the equation, $\ln(S)$ is mean reverting to $Ln(m)$. We designate this type of mean reversion process MRL.

All three processes described thus far (GBM, MRV, and MRL) are smooth processes. There are no big moves from one time step to another. They do not "jump." However, it is also possible to introduce jumps into these processes. The events on September 11, 2001 created a catastrophic move in prices of assets that will not happen in the type of smooth processes described thus far. We can accommodate this by adding a jump term in the stochastic process:

$$dS_t = \mu S_t dt + \sigma S_t dW_t + dq$$

$$dq = \phi(p, pud, N(m, sd))$$

where dq represents a jump in dt with certain probability p and a jump magnitude of a sample from a normal density function with average m and standard deviation sd and with probability pud that the jump is up (positive).

Similarly, we can introduce jumps in the mean reverting process:

$$\ln(dS_t) = \eta(\ln(m) - \ln(S_t))dt + \sigma dW_t + dq$$

or

$$S_t = \eta(m - S_t)dt + \sigma dW_t + dq$$

In these cases, the jump will be neutralized over time by the mean reversion effects. After the jump, prices will be pulled back to the long-run mean faster as they will instantaneously deviate from the long-run average significantly (more than what could be expected from a smooth process) and exert a pressure opposite to the direction of deviation (for reasons previously explained).

It is also possible that when a jump occurs, the process fundamentally changes and assumes a different reversion rate and possibly a different long-run average mean. One could imagine significant technology change resulting in a jump and simultaneously settling into a different price process with a new reversion rate and long-run average mean. We call these *regimes*. The price process may be in regime 1 with certain characteristics until it is met with a "shock" resulting in a jump that kicks the process to another regime 2. In the new regime, the price process characteristics can be very different from the preceding regime. If this is a tactical shift, yet another shock can reset the process back to the original regime. If the jump represents a permanent change, the process does not reset and continues in the current regime until another shock puts it in yet another new regime. For example, the shock related to the 9/11 events may have been a tactical regime shift for many assets—most of which reset back to long-run characteristics soon afterward. However, the financial shock related to the credit crunch and real estate collapse may have long-run effects on the U.S. economy. Regime shifts related to this shock may never reset, and this may have implications for the long-run growth and savings rate in the U.S. economy.

To summarize, if the weak form of EMH holds (as is the case in most empirical tests), we can model the price processes of traded assets using GBM as long as the asset does not have mean reversion. Mean reversion happens when the supply and demand of the asset (as in the case of commodities) respond to the current prices. As prices move higher, supply increases (due to increased investments) and demand declines (aided by conservation), forcing prices down (reversion to the mean). Similarly, when prices move lower, supply decreases (due to reduced investments) and demand increases (as conservation decreases), resulting in an upward thrust on prices. Without further rigor, we assume that all commodities are modeled using the mean reverting process. It should be noted that if the rate of reversion is low, the process will be very similar to the GBM, and either will give reasonable results.

6

Self-Standing Options and Pricing of Options

An option is a right but not an obligation to do something. A European call option on a stock in the financial market gives the holder the right to buy a share of stock (*exercise the option*) at a predetermined price (*strike price*) at a predetermined time (*time of expiry*). The holder of such an option is under no obligation to buy the share of stock or do anything for that matter. But, if the price of the stock is higher than the strike price when the option can be exercised, the holder of the option most likely will exercise it by paying the strike price for a share of the stock. In effect, the holder of the option can pocket the difference between the stock price and the strike price. But, if the stock price is lower than the strike price, the holder of the option will not exercise it and will let the option expire worthless. An option is valuable to its owner as the worst that could happen is that the option expires worthless. Since the price of stock does not have a technical upper bound, this difference between the stock price and strike price also does not have an upper bound (although the price of the stock is unlikely to exceed certain limits based on the characteristics of the industry and overall economy).

How can one price an option when it is bought or sold? Clearly, the value of the option depends very much on the price of the stock on which it is based. Consider a call option, which gives the right to purchase a share of stock at a future time for a prespecified strike price. The higher the stock price at the time of purchase of the call option, the higher the value of the call option will be. Similarly, the lower the strike price, the higher the value of the call option will be. Intuitively, it may also be clear that the more volatile the stock is, the more valuable the option is as well. This is because the "payoff" of an option is asymmetric. If the price is lower than the strike price at the time of exercise, the gain is zero. The gain is always zero regardless of how much lower the price is compared to the strike price—as long as the price is less, it does not matter. On the other hand, if the price is higher than the strike price, the gain is proportionately higher. Another factor that affects the value of the option is the time to exercise. The longer the time to exercise, the broader the range of possible outcomes of the stock price at the time of exercise. As such, the option value is also higher when the time to exercise is longer. Another factor that also affects the value of the option is the interest rate, the rate at which money can be borrowed. Why interest rates affect option pricing will become clear once we have studied the option pricing techniques described in this chapter.

To understand risk-neutral pricing of options, consider the following stylized example: Suppose we have a 1-year call option on stock XYZ at a strike price of $110. The current price of the stock is $100. Further assume that stock XYZ can only take two values in 1 year: It will either drop to $80 or rise to $125. This is a measure of *volatility* of XYZ, its tendency to go up or down. Although in real life the stock can take many different possible values, the problem is simplified in this stylized case. It can be shown that if the time step is made smaller and smaller, at the limit we can safely assume that the stock price moves to only two possible values. Call the state in which the price drops to $80 the "bad state" and the state in which the price moves up to $125 the "good state." Since the call option has a strike price of $110, it will have a value of $0 in the bad state and a value of $15 ($125 – $110) in the good state. Remember that the call option gives the owner the right but not an obligation to buy one share of XYZ at $110. So, the option will mature worthless in the bad state, and it will be exercised for a profit of $15 in the good state.

Also assume that the interest rate is 10%. We can now create a *synthetic option*, something that will have the same payoff in the bad and good states. We do this by creating a bundle that has one share of stock and $72.70 in debt (we borrow $72.70 at an interest of 10% per year). This bundle currently is worth $27.3 ($100 [current price of stock] – $72.70 [borrowed money] = $27.3). Think of this bundle as a security that has two components (which always go together). In this case, the two components are one share of stock and the obligation to pay back $72.70.

Look at what this synthetic option will be worth in the future. In the good state, the price of the stock moves to $125, and we have to pay back the borrowed money with interest ($72.70 + $7.30 = $80), so the synthetic option will be worth $45 ($125 – $80). In the bad state, the stock moves to $80, and we owe the bank $80 ($72.70 + $7.30), so the synthetic option is worth $0. This is very convenient as the synthetic option that we created with a share of the stock and borrowed money has the same payoff as three times the original option. Remember that the original option had a value of $15 in the good state and $0 in the bad state. If the synthetic option is three times as potent as the actual option, its current value also will be three times the value of the original option. If not, one can make arbitrage (no-risk) profits.

For example, if the synthetic option (stock + loan) is priced higher than the price of three options, one can sell (*short*) the synthetic option bundle and buy (*long*) three options. At maturity, this position will yield a profit regardless of the state of the economy. Similarly, if the synthetic option is priced lower, one can enter into a reverse trade by going long on the synthetic option bundle and short (*write*) the options and make guaranteed profit regardless of the state of the economy. In efficient markets, such arbitrage profits (profits that carry no risk) cannot exist. We established that the synthetic option was worth $27.3 today; hence, we can conclude that the original option is worth $27.3/3 = $9.10 as it has a payoff of ⅓ of the synthetic option bundle.

$$OptionValue = \Delta \times (StockPrice - Loan)$$

We call Δ the number of synthetic option bundles needed to pay exactly the option. In the example, Δ is ⅓.

$$OptionValue = 1/3 \times (100 - 72.7) = 9.1$$

Why is this important? This simple stylized example gives us an elegant way to price options. We priced an option by creating a synthetic option (which included the underlying stock) and invoked a no arbitrage argument (i.e., bundles of securities with the same future payoff characteristics should have the same price today). This is irrespective of the risk preferences of the investors, so we can now take an elegant leap forward and say that the mechanics of the options pricing will work under any risk. For example, we can price the option in a risk-free world, and we can safely ignore the risk characteristics of the stock.

Let us revisit the first example. Now assume that we are in a risk-free world where the risk-free rate is 10%. We can borrow and lend money at the risk-free rate of 10%. The risk-free rate is observable in the marketplace. Typically, we use the yield on 90-day T-bills as a proxy for the risk-free rate as the chance of the U.S. government going bankrupt in such a short period of time is virtually zero.

In such a world, every asset will return exactly the same—the risk-free rate (we assumed this to be 10%). Remember that the stock dropped to $80 in the bad state and rose to $125 in the good state. So, the return in the bad state is −20% and the return in the good state is 25%.

Since the expected return on the stock (and every other asset) in this world is 10%, we can calculate the probability of the bad and good states occurring. Say that p is the probability of the good state. This means that the probability of the bad state will be $(1 - p)$ as there are only two states.

$$10\% = p \times 25\% + (1 - p) \times -20\%$$

$$p = 67\%$$

This means that there is a ⅔ chance that the future state will be good and a ⅓ chance that the future state will be bad. Since the price of the option is $15 in the good state and $0 in the bad, the expected value of the option is $15 × 67% = $10. Since the return on all assets is the risk-free rate of 10%, today's value of the option is 10/1.1 = $9.10, the same as the prior calculation.

This simple example demonstrates that as long as we can create a synthetic bundle by combining the underlying asset with a loan that has the same pay-off as the option on the asset, we can price the option without consideration for risk characteristics. We can simply assume that we are in a risk-neutral world and price the option using the risk-free rate. In the stylized example, we assumed that there are only two possible outcomes, but this obviously is not the case for a real asset. We can now combine the idea of risk-neutral pricing with the stochastic processes discussed.

In the discussion, we established that the price process of traded assets can be assumed to follow geometric Brownian motion (GBM). This assumption satisfies the efficient market hypothesis. GBM in a risk-neutral world will have a risk-free drift. Remember that the GBM has two terms: a drift term and a volatility term. The price progression of an asset (such as stock) will have a drift commensurate with the expected return on the stock. The expected return on any asset in the risk-free world is the risk-free rate; hence, the drift of the price process followed also will be the risk-free rate.

Remember that the expected return is a function of the β (or systematic risk) of the asset. Hence, the real drift of the price process followed by an asset will be its expected return. However, to value an option on such an asset, we can deploy the risk-neutral valuation technique. We do this by providing a risk-neutral drift to the price process followed by the asset (assumed to be GBM). This is analogous to finding the probability of the good and bad states in the two-state stylized example.

Figure 6.1 shows the simulation of the price process of HPQ (Hewlett Packard Company) with a risk-free rate of 2% for a period of 1 year starting

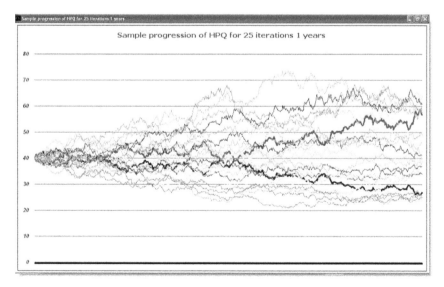

FIGURE 6.1
Stochastic simulation of HPQ price.

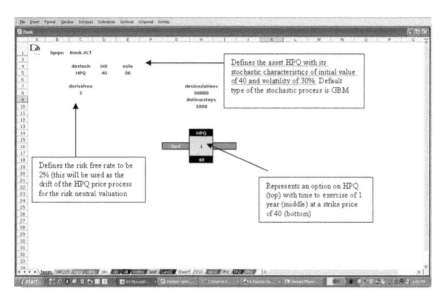

FIGURE 6.2
Decision Options framework for the representation of an option.

with an initial price of $40 and volatility of 30%. Note that the volatility term has significant influence over the progression of the price process; thus, the actual drift of the simulated price process is difficult to ascertain just by observation. In some sense, it gets lost in the "noise." In any case, the risk-neutral simulation of the HPQ price is the input we can use to price an option on the stock. Assume that we are interested in pricing a call option on HPQ with the following characteristics: European option type (we can only exercise at a specific time), 1 year time to exercise, and $40 strike price.

At this point, I would like to introduce the proprietary software called Decision Options Technology (DoT), which can be used to solve complex options problems. Figure 6.2 is a representation of the HPQ call option within DoT. The framework shown has the following constructs:

- A pictorial representation of the option that shows its fundamental characteristics: name of the asset, time to exercise, and strike price. The option is shown as a block with three bands; the asset is shown on the top of the block, the time in the middle, and the strike price at the bottom. The two gray handles on either side are for connecting it to other entities. The name of the option is provided on the handle to the left. In this case, it is called Do1. The underlying asset for this option is the stock price of HPQ, represented as HPQ in the top band. The time to exercise is 1 year, as shown in the middle. The strike price is $40, as shown in the bottom band.

- The representation of the asset (in this case, a share of the HPQ stock) with its fundamental characteristics: the type of price process, initial value, and its volatility. The term *dostoch* tells the program that the asset HPQ follows a stochastic process. No definition of the stochastic process is provided, so the program assumes that HPQ follows GBM. The initial value is the current price of HPQ, which can be observed in the marketplace, and volatility is a measure of how much HPQ price bounces around over time. The initial price (today's price) is $40 (identified as "init"), and the volatility is given as 30% (identified as "vola").

- The representation of the risk-free rate (in this case, we assumed 2%). This is represented as "doriskfree."

- Also provided are simulation parameters. The term *dosimulations* represents how many simulations are to be run, and the term *dotimesteps* represents how many time steps are used in each simulation. By referring to the picture of a few sample risk-free simulations of HPQ (see Figure 6.3), you can see that each simulation (represented by a single line starting today and ending 1 year from now) is discretized into a large number of time steps. By using 1,000 time steps, we discretize the stochastic progression of HPQ into 1,000 steps, each representing approximately 36% of a day (365/1,000).

The definition of the option is now complete, and we can price this option using risk-neutral stochastic simulation. Note that this problem can be easily solved using a Black-Scholes equation, which is a closed-form solution for such single-standing financial options as this one. We return to why simulation is a useful technique for options pricing as well as how to use the Black-Scholes equation later in this chapter.

The 1-year option on HPQ stock at a strike price of $40 is priced at $5.20. If such an option traded in the marketplace, we expect it to trade at approximately $5.20. Options markets are not as liquid as the stock market, and the actual price may depend on liquidity constraints, but we have established that the efficient price of such an option is $5.20. If the option is priced higher or lower, it will represent arbitrage profits that will be quickly neutralized in efficient markets. With computers scanning the markets continuously looking for such price discrepancies and eager to make markets "efficient" by taking away the arbitrage profits, it is difficult to find mispricing in large public markets.

How do we calculate the option price using simulation? First, we simulate the price of HPQ stock in a risk-neutral world with a risk-free drift (as discussed). For each simulation of the price process, we get a price at the end of 1 year. If this price is above the strike price of the option (in this case $40), we exercise the option and get a value that is the difference between the stock

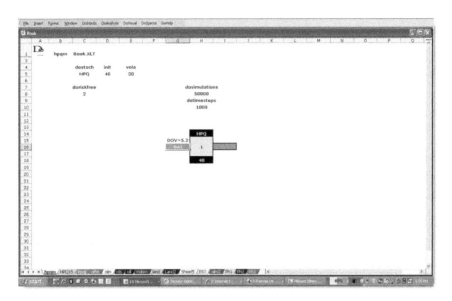

FIGURE 6.3
Pricing of HPQ call option using Decision Options Technology.

price at the end of 1 year and the strike price. We then discount this value using the risk-free rate back to today, which gives us the value of the option in that simulation. We then repeat this process thousands of times (50,000 times in this case), and in each simulation we get a value for the option. We then take an expectation of all these values (by averaging them), which gives us the value of the option today.

We can visualize the risk-neutral payoff diagram in these simulations. Figure 6.4 shows the risk-neutral payoff distribution from the HPQ call option for 1 million simulations. The x axis shows the value of the option, and the y axis shows the probability of getting that value in a simulation. This is similar to frequency diagrams or histograms in which the value is shown on the x axis and frequency is shown on the y axis.

Note that this is the risk-neutral payoff and does not represent the actual payoff from the option. We solved the option problem in the risk-neutral world (using the replication and no arbitrage technique explained in this chapter) so that we do not have to worry about the actual risk of the stock and the option. Note that HPQ stock will have a risk commensurate with the systematic risk carried by Hewlett-Packard equity (which is a function of its operating risk and financial leverage), which is certainly different from a risk-free rate. More important, the option on HPQ that will have significantly higher volatility (depending on the price of HPQ and the time to expiry) will carry systematic risk many-fold higher than HPQ itself. The systematic

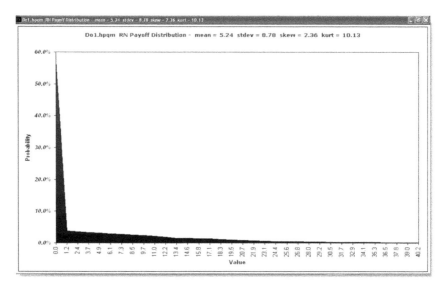

FIGURE 6.4
Risk-neutral payoff from HPQ call option.

risk of the option is nearly impossible to determine as it has a complex and asymmetric payoff in different states of the world.

However, the risk-neutral payoff is instructive in that we can get a feel for the probability of making money from buying the option. Figure 6.4 shows that approximately 50% of the time the option will expire worthless (this will happen if the HPQ price 1 year from now is at or less than 40). The remaining 50% of the time, we will exercise and make some money from the option. A small percentage of the time, we will make large amounts of money, and the return from the option will be huge. Remember that the option is currently priced at approximately $5, so anything above $5 is money in the pocket, so a $10 payoff is a 100% return in 1 year. Be cautioned not to attach "real meaning" to these numbers as these are from the risk-neutral simulation. In the real world, the actual payoff distribution will be slightly different. We are not assuming that HPQ has risk-free characteristics. We are just using the risk-neutral technique to value the option on HPQ.

Now, also look at what impact the parameters of the stochastic process defining HPQ (its initial value and volatility) have on the option value. The two parameters that affect option value are the initial price and the volatility of the underlying asset (HPQ stock), assuming that everything else such as the strike price, time to expiry, and interest rate remain the same. We can run an impact analysis to accomplish this. To assess the impact of one parameter, say volatility, we keep everything else the same and change volatility (a bit higher and a bit lower) and value the option. Figure 6.5 shows that a 1% increase in volatility (from the assumed 30%) increases option value by 0.9%. It is nearly symmetrical on the downside; that is, a 1% reduction in volatility

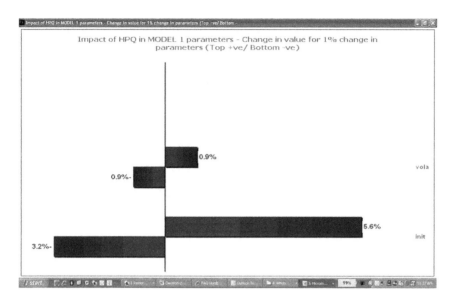

FIGURE 6.5
Impact analysis on options price.

reduces option value by 0.9%. This allows us to understand how the valuation will change if we were slightly off in our calculation of volatility. We do the same exercise for initial value. For a financial option, the current price can be observed in the marketplace, so it may not make intuitive sense to do an impact analysis. As we discuss in this book, for real assets our estimate of initial value of the underlying asset is an estimate as in many cases we cannot observe it in the market directly. So, understanding how the valuation changes as the assumption of initial value changes will help us assess the robustness of our conclusions and decisions. In this case, we find that the option value increases by nearly 6% by a 1% increase in initial value. On the downside (i.e., if the initial value is less by 1%), the value drops only 3% (less than in the case on the positive side). Remember that the downside risk is eliminated for options. If the option matures out of the money, it does not really matter how far below the strike price the stock price ends. The option will not be exercised, and the payoff will be zero. On the upside, however, it does make a difference how much more the stock price is as the gain from options exercising is the difference between the stock price and the strike price.

Note that we obtain these results from simulation, so a large number of simulations and time steps may be necessary to get precise values. Also note that the "impact" of a parameter on option value is not linear. So, the results above make sense only within certain ranges of changes. For most cases involving decision options on real assets, we have an idea how far off we may be on the assumptions (in other words, we may have optimistic and

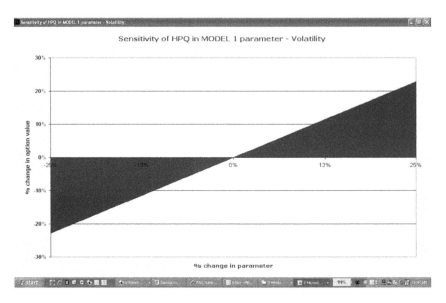

FIGURE 6.6
Sensitivity analysis of volatility.

pessimistic scenarios that will bound the limits of input variability). To better understand how the option value changes as the value of the stochastic parameter changes, we can run a sensitivity analysis. For example, Figure 6.6 shows the results of the sensitivity analysis of volatility.

In the previous problem, we assumed that the price of HPQ follows plain vanilla GBM. Now let us introduce some jumps into the process. Assume that there is a 50% probability that a jump will happen before the maturity of the option. If a jump happens, the stock loses 20% of its value. We also assume that in 90% of the cases, such shocks are negative (price drops). In 10% of the remaining cases, the stock rises by an equal amount (20% of the price just prior to the jump event). One could imagine such price moves, such as when the company disappoints in earnings due to unforeseen events from what the market expected. Note that these are simplified assumptions of more real-world cases as it is impossible to precisely determine either the probability of a jump or the magnitude of the jump as these occur due to unforeseen events. As we will see, we can estimate certain characteristics of such jumps based on historical data, especially if the characteristics are driven by repeating events.

Some have pointed out the erratic behavior of stock prices at the release of financial results of the company as evidence that markets are not efficient. Academic studies have shown that such market adjustments to release of financial statements are complex as they have to take into account changes from financial expectations that existed prior to the release of the information, possible signaling effect by management, and possible discounting of

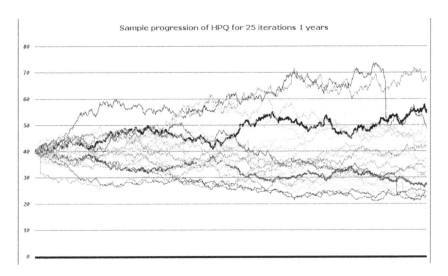

FIGURE 6.7
Stochastic simulation of HPQ price with jumps.

the market's expectation of management competency that exists at the company. Remember that if markets are weak-form efficient but not semistrong efficient, the prices can (and will) change as the company releases information unknown to the market. A jump process (an instantaneous change in price as new information is released) confirms that the market behaves as expected. Further analysis of the company's financial analysis may not help make excess profits as the change is instantaneous.

We can visualize the HPQ price process with jumps as shown in Figure 6.7. In this case, we assumed that there is a 50% chance that the jump will happen in 1 year, and if the event happens the price will move 20% up or down from the price just prior to the occurrence of the event. The jumps are also biased to the downside by providing only a 10% chance of a positive jump. You can see significant jumpy moves of the price in certain simulations.

Such a process that has a much higher probability of a negative shock to the asset price will decrease the value of the option. Let us price the 1-year HPQ option discussed. The possibility of the asymmetric negative shock substantially reduces the value of the option (from $5.20 to $3.50; Figure 6.8).

Let us also briefly discuss a closed-form solution for self-standing options called the Black-Scholes equation. The Black-Scholes equation can be used to price options when the underlying stochastic process is GBM and when options have no interactions. This means that it is valid only for single options (not combinations of options that may interact). It also cannot be used when the stochastic process is not GBM and may show mean reversion characteristics, as in the case of commodities. It also cannot be used when there are jumps in the stochastic process of the underlying asset.

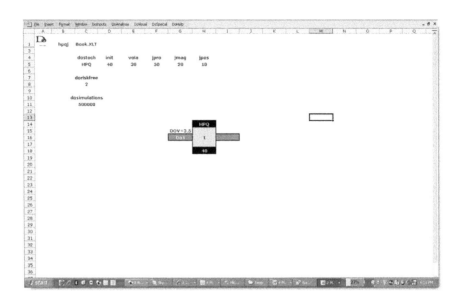

FIGURE 6.8
Pricing of HPQ option with negative jumps in HPQ price process.

The Black-Scholes equation is as follows:

$$C = S\phi(d1) - Ke^{-rT}\phi(d2)$$

$$d1 = \frac{\ln(S/K) + (r + \sigma^2)T}{\sigma\sqrt{T}}$$

$$d2 = d1 - \sigma\sqrt{T}$$

where
  C = Call price
  S = Stock price
  K = Strike price
  T = Time to exercise
$\phi(d1)$ = Cumulative standard normal distribution ($N(0,1)$) = delta

Although the equation is somewhat intimidating, there is a simpler way to understand it. Remember our stylized risk-neutral valuation of the hypothetical asset. We priced it by creating a synthetic option containing the underlying stock and a loan. The Black-Scholes equation is a representation of the process we used. The first term represents the stock and the Δ (the number of synthetic options needed to replicate the option payoff exactly). So, this is the current value of the number of stocks in our synthetic option bundle. In our stylized example, the Δ was ⅓.

The second term represents the loan, which is the present value of the amount we need to borrow to create the synthetic option bundle. Note that the loan is a function of the present value of the exercise price. The higher the exercise price, the more money we need to borrow to create the synthetic option. The current value of the option then is the difference between the value of stock in the bundle and the money borrowed, essentially the value of the synthetic bundle, multiplied by Δ. Just to understand how the mechanics works, let us price the HPQ option we solved using simulation.

For the HPQ example, we have the following inputs: $S = 40$, $K = 40$, $T = 1$, $r = 2\%$, and $\sigma = 30\%$

$$d1 = \frac{\ln(40/40)+(2\%+30\%^2)1}{30\%\sqrt{1}} = 0.006$$

$$d2 = 0.006 - 30\%\sqrt{1} = -0.294$$

$\phi(d1)$ = Probability that a sample from $N(0,1)$ will be less than $d1$

In other words, it is the cumulative probability of $d1$ in a standard normal function. Figure 6.9 indicates the cumulative probability distribution of a standard normal function. For $d1 = 0.006$, we can calculate $\phi(d1) = 0.55$ (Δ). For $d2 = -0.294$, we can calculate $\phi(d2) = 0.43$.

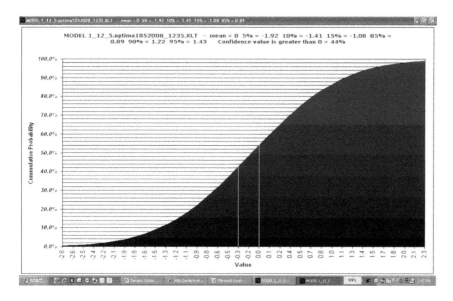

FIGURE 6.9
Standard cumulative probability distribution for Black-Scholes parameters.

$$C = S\phi(d1) - Ke^{-rT}\phi(d2)$$

$$C = 40 \times 0.55 - 40 \times e^{-0.02} \times 0.4 = 5.14$$

The value of the option according to the Black-Scholes equation is $5.14 and is close to the value we calculated using the risk-neutral simulation method ($5.20).

Let us also introduce another concept at this stage: the dividend. As intuition will tell us if the stock paid a dividend, its price will decline (as cash is returned to shareholders). So, an option on a dividend paying stock will be slightly less than one that does not pay dividends, keeping everything else constant. Let us return to the HPQ example and assume that Hewlett-Packard pays a dividend of 2% per year. We can incorporate this into the stochastic process by giving it a negative drift of 2%. Remember that in the risk-neutral world, the expected drift is the risk-free rate, so the negative drift will be applied after the positive drift of the risk-free rate. So, if the risk-free rate is more than the dividend rate, the stochastic process will still have a positive drift.

Mathematically, we can modify the differential equation for GBM by introducing the dividend δ as a modification to the drift term. In the risk-neutral world, μ is the risk-free rate, and the actual drift of the stochastic price process will then be $(\mu - \delta)$, which will be positive if the risk-free rate is higher than the dividend and negative otherwise.

$$dS_t = (\mu - \delta)S_t dt + \sigma S_t dW_t$$

Figure 6.10 gives the price of the option for HPQ when there is a 2% dividend yield (but not the jumps discussed previously). The option price decreases to $4.70 from the original $5.20 if the underlying stock pays a 2% annual dividend. The 2% dividend will result in $0.80 (based on the initial price) to be returned to shareholders and correspondingly reduce the stock price by that much. Note that the value of the option has not gone down by the amount of the dividend. This is because of the asymmetric payoff of options. If the option matures out of the money, it does not really matter whether the company issues a dividend. The option will not be exercised.

We can do a similar exercise for a commodity. To make a similar comparison, assume that we have an option to buy one barrel of oil at $100 a year from now. The current price of oil is $100/barrel with a volatility of 20%. The half-life of oil is 2 years, and the long-run average expectation of oil prices is $60/barrel. Current risk-free rate is 2%. What is the price of this option today?

We can price this option using the same techniques as the HPQ example. We can simulate the price process of oil in a risk-neutral world (Figure 6.11). In this case, the underlying asset is the oil itself. Because it is a commodity driven by supply-and-demand pressures (which are a function of the current

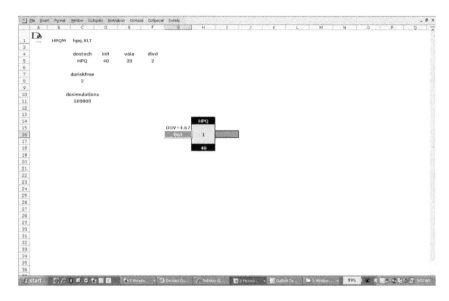

FIGURE 6.10
Valuation of HPQ option with dividend yield.

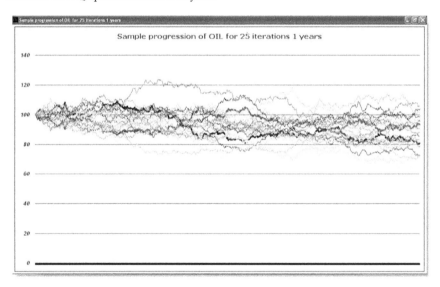

FIGURE 6.11
Stochastic simulation of oil price.

price), it exhibits mean reversion to a long-run average. In Figure 6.11, we can see the effect of mean reversion in oil price. Since the half-life is assumed to be 2 years, it does take significant time for prices to revert to long-run average after a big excursion from it. The current price of $100 is significantly different from the long-run average price of $60.

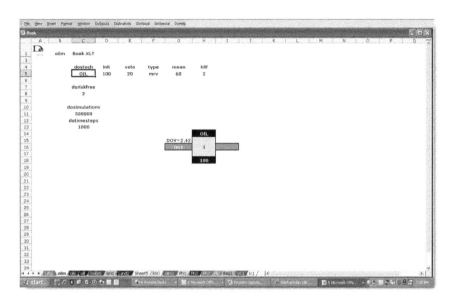

FIGURE 6.12
Pricing of option to buy oil.

As shown in Figure 6.12, this option is priced at approximately $2.40. Note that even though the long-run average price is a lot lower than the strike price, the current high price coupled with a half-life of 2 years results in a relatively high option value. This means that one is willing to pay up to $2.40 for the right to buy one barrel of oil 1 year from now for $100.

In this example, we assumed that the price of oil mean reverts to the long-run average (the stochastic MRV process). If we instead use the $\ln(P)$ as the mean reverting variable (the stochastic process MRL), we get a slightly lower value for this option ($1.60), as shown in Figure 6.13.

So, the assumptions on the stochastic characteristics of the oil price are quite important in pricing this option. The initial price is observed in the marketplace, and the volatility can be easily derived from historical prices. Derivation of half-life and long-run average mean values is more complicated, and there are a number of numerical techniques available to do so. As in any valuation, it is always important to assess the sensitivity of assumptions to ensure the robustness of the valuation and conclusions made.

Commodities do not have dividends, but there is a similar concept called *convenience yield*. The discussion around convenience yields can get a bit theoretical, but it is sufficient to remember that convenience yield for a commodity happens because of the "convenience" of owning the physical commodity rather than a long futures contract on the commodity. You can imagine this as an option the holder has on the physical commodity; the holder can either consume it today or store it for the future. This "timing option" of when to consume is not available to those who are holding only

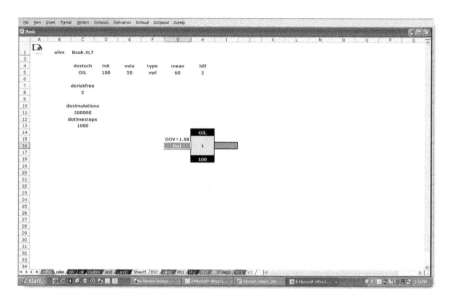

FIGURE 6.13
Oil option with mean reversion in ln(P).

futures contracts. They do not have the commodity (yet); they only have a contract to take delivery of the commodity at a future time. The holder of a physical commodity incurs storage costs. So, the convenience yield can be positive or negative.

When the convenience yield is positive, the futures price of the commodity will be lower than the spot price. This is called *backwardation*. This is typically the case for most commodities. In the case of crude oil, more than 75% of the time a futures curve shows backwardation.

We can now modify our mean reverting stochastic equation by incorporating convenience yield to "correct" the discount rate. Remember that the original equation is

$$dS_t = \eta(m - S_t)dt + \sigma dW_t$$

The real drift of this mean reverting process is $\eta(m - S_t)dt$, where η is the reversion rate, and m is the long-run average mean. Assume that the risk-adjusted discount rate is μ as we used previously. As established, the convenience yield δ will be a correction to the discount rate. Mathematically,

$$\delta = \mu - \eta(m - S_t)$$

To price options, however, we need to simulate a risk-neutral mean reverting process. If the risk-free rate is r, we can then equate

$$r - \delta = r - \mu + \eta(m - S_t)$$

This allows us to modify the drift term in the mean reverting stochastic process in a risk-neutral world as

$$dS_t = (r - \mu + \eta(m - S_t))dt + \sigma dt W_t$$

Practically, in the risk-neutral world, there is a correction that needs to be applied to the drift term. We subtract $\mu - r$, where μ is the discount rate of the underlying commodity.

Commodities are driven by both demand and supply and thus exhibit low correlation with the broad market. Hence, we will assume that when we deal with commodities, the systematic risk of the underlying commodity is close to zero, and the discount rate is the risk-free rate. The correction described becomes zero, and we can simulate the price process without a correction for the risk-neutral valuation. We thus use a real risk-free discount rate of 0% for all commodities in this book.

Let us also value the option with positive shocks. Assume that within 1 year, there is a 20% probability of a superspike in oil prices. The superspike doubles oil prices in a very short period of time. The reversion rate (and hence the half-life) and long-run average mean of the price process do not change, however. In the real world, a superspike may represent a significant change in the demand-and-supply characteristics of oil, and most likely many of the primary parameters of the process such as volatility, long-run mean, and half-life need to be modified. In this case, we assume that all of those parameters remain the same pre- and postspike. Figure 6.14 indicates the price process with some scenarios showing the effect of superspikes and the price of the option in the presence of superspikes. Note that mean reversion is not a guarantee that prices will move back, and there can be situations when the superspike is followed by a continued move up. However, the further the price is from the long-run mean, the higher the force is to bring it down in subsequent periods.

The 1-year option is of significant value in this case ($19), as shown in Figure 6.15. A 20% chance of a superspike (defined as the doubling of price) increases the value of the option many-fold.

In summary, we can price individual options on a single asset, such as a call option on a stock, by simulating the price process of the stock in the risk-neutral world and taking the present value of the expectation of value at exercise. For assets that follow GBM, the Black-Scholes equation allows a closed-form solution to reach the same answer. However, the Black-Scholes equation cannot be applied when the price process of the underlying asset is not GBM or if the option is not single standing. In all cases, we can value an option on the underlying asset by risk-neutral valuation. We use assumptions of replication (creation of a synthetic

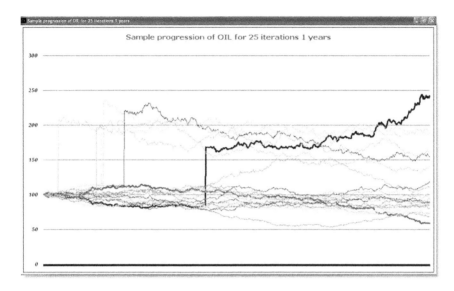

FIGURE 6.14
Stochastic simulation of oil price with superspikes.

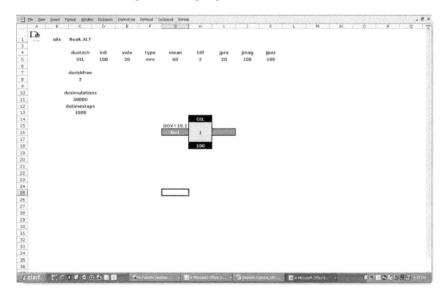

FIGURE 6.15
Pricing of oil option with superspikes in oil price.

bundle with the same payoff as the option) and no arbitrage to create a
framework of risk-neutral valuation of options. When dividend or convenience yield is present, the risk-neutral stochastic drift can be modified
to account for it.

Commodities generally exhibit mean reversion and do not follow GBM. Options on commodities can also be solved by simulating the mean reverting stochastic process in a risk-neutral framework. Since the correlation between commodities and market is low, we generally assume that the discount rate on the underlying is the risk-free rate.

7

Pricing of Decision Options

We are now ready to tackle decision options—a general problem that contains multiple interacting options, cash flows, private risks, and other entities. Options pricing theory was invented to price financial options. As we have seen, financial options are single-standing. When they are on assets such as stocks, whose price process follows geometric Brownian motion (GBM), we have a very convenient closed-form solution in the Black-Scholes equation. Both risk-neutral valuation techniques and the Black-Scholes equation rely on the assumption that we can invoke a no-arbitrage condition against a synthetic bundle, formed using the underlying asset and debt, that has the same payoff as the option and the option itself. As we discussed, in an efficient market, assets with the same future payoffs will be priced the same today.

In all the examples we considered thus far, we had only a single option on a single underlying asset. In the examples that follow, we analyze complex option bundles that interact with each other. Throughout the book, we make liberal use of the software Decision Options Technology (DoT). DoT utilizes constructs that are predesigned to represent various types of decisions (options), private risks, and cash flows. It also has a schema to describe underlying assumptions as well as reporting of results.

DoT analysis typically shows three areas. It shows the assumptions for the model on top, a graphical representation of the decision problem in the middle, and the entire valuation results at the bottom (Figure 7.1). The assumptions are generally divided into different areas. They largely fall into the following three types: stochastics (time series), probabilities or constants, and simulation parameters.

As explained, stochastics are used to represent tradable assets that are driven by market risks. Probabilities or constants are used to represent private risks, predetermined outcomes, or time. Simulation parameters include number of simulations, time steps, discount rate, risk-free rate, and tax rate (Figure 7.2).

The model is a graphical representation of the decision problem being solved as shown in Figure 7.3. Each of the boxes in the picture represents a decision or private risk. A decision can be an option, cash flow (a predetermined outcome), and as we see later, a swap or option to swap. As you have seen, the box representing a private risk is characterized by shading in the middle.

Information is provided on top, in the middle, and at the bottom of the box. In the case of decisions (options, cash flows, and swaps), the top represents

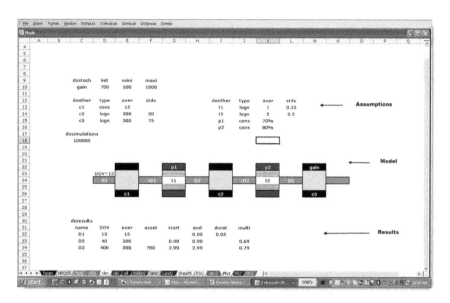

FIGURE 7.1
Decision Options Technology (DoT) software schema.

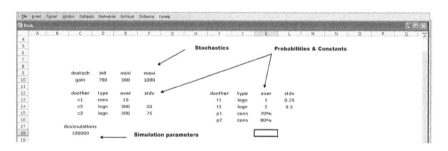

FIGURE 7.2
Decision Options input schema.

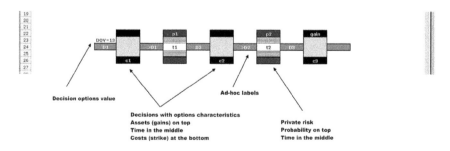

FIGURE 7.3
Decision Options model schema.

the asset (what you get by making a decision or in the case of a cash flow what is given to you), the middle represents the time available to make a decision or receive a cash flow (in all examples described here we use years as the unit of time), and the bottom represents a cost (on which you have an option to pay—the strike price—in the case of options and swap and what you have to pay in the case of a cash flow). For private risks (as represented by boxes with the shading in the middle), the top represents a risk, and the middle represents time in years. The risk can be a discrete probability or a probability distribution. The distributions can be of any type, including a binary outcome that takes a value of 1.0 or 0.0 as in the case of life sciences experiments that reveal information that results in a complete abandonment of a program due to unacceptable drug toxicity.

A decision sequence moves from the left to right. The total duration of the decision problem is the sum of all the time (in years) shown in the middle of all boxes (whether decisions or private risks). The time in each box represents the available time from the end of the previous decision or risk to the one being considered.

In some cases, models may also include a decision tree-like construct as shown in Figure 7.4. These constructs are not decision trees but are decision option trees. Some of them may be "OR" or "AND" trees that allow decision flexibility in decisions, and some of them may take probabilities or probability distributions and will act like private risks. So, the tree constructs can represent optionality or just private risks.

In combination, this framework and constructs of the software provide complete modeling functionality of most known decision problems with many different types of uncertainties and decision flexibilities. They also allow users to switch off "uncertainty" or "flexibility" to better understand how these attributes affect their decisions. It should be noted that switching off flexibility in a decision options tree will make it act like traditional decision trees with uncertainty and will allow the user to create probability distributions of discounted cash flow (DCF) net present value (NPV) using traditional

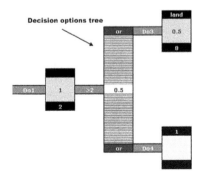

FIGURE 7.4
Decision Options tree schema.

FIGURE 7.5
Decision Options representation of combination option.

Monte Carlo simulation. By switching off uncertainty, decision options trees will behave as traditional decision trees in decision tree analysis (DTA), and if the tree structure is not present, they will reduce to traditional DCF analysis. Thus, this framework allows the user to learn the effects of constraining assumptions in traditional methodologies as well as to make better decisions considering all available information: uncertainty and flexibility.

Now let us consider a new type of security called MSIBM. This is a combination option. It gives you an option to buy one share of IBM in 1 year at a strike price of $120. If you do exercise that option, 1 year later you can make another option exercise decision. This time, you have the option to exchange an IBM share for four shares of Microsoft.

IBM's price today is $120 and shows a volatility of 20%. MSFT's price today is $30 and shows a volatility of 25%. The risk-free rate is a bargain 2%. What is the value of this option today? Figure 7.5 is a representation of this decision option. The two blocks in the diagram represent the two options. These two options are sequential; that is, the holder of this instrument has to make two exercise decisions in sequence. At the end of the first year, the holder has to make a decision to exercise the first option. If the holder does not exercise the first option (to buy one share of IBM at $120), the holder forfeits not only the first option but also the second one. This means that the exercise of the first option is necessary to keep the second option alive. As you can appreciate, this makes the first exercise decision (in 1 year) a complex one. As we have seen, for single-standing options, exercise decisions are trivial; at the time of expiry, the option holder only has to observe the market price of the asset and compare it against the strike price. If the asset price is higher than the strike price, the holder will exercise and otherwise not. In the case of this combination option, however, the first exercise decision needs to take into account the value of keeping the second option alive. In other words, the asset that is "bought" by the first exercise decision is not just one share of IBM but an additional option to exchange one share of IBM for four shares of MSFT 1 year later. If you are beginning to feel overwhelmed, I do not blame you; it is not a simple decision.

Figure 7.5 is the schema representing the combination option (decision options). The boxes represent the two options and the text on top of the figure describes the price processes associated with the assets (in this case the stock of MSFT and IBM). In the boxes, the top band shows the asset (the stock), and the bottom band shows the strike price. In the middle, the time to expiry is also shown.

This problem cannot be solved using conventional methods or the Black-Scholes equation. In pricing the option combination today, we have to consider not only what will happen 1 year from now but also what will happen 2 years from now and how that will affect your decision 1 year from now. The option to buy one share of IBM 1 year from now is not simply an option on an underlying asset, but it is an option on an option as well. These are *interacting options*. The value of this combination option is driven by two underlying assets: IBM and MSFT. As can be seen from Figure 7.5, we have defined the initial prices of IBM and MSFT to be $120 and $30, respectively (these are observed in the marketplace today). We have also defined the volatility of IBM and MSFT to be 20% and 25%, respectively (these were calculated from historical prices of the two stocks).

Assume that we priced this option combination using some technique today, and we decided to buy it. One year from now, we will conduct an analysis to see if we should exercise the first option. Note that if it were a single option, this decision will be a trivial one; if the price of IBM is higher than $120, we will exercise, and we will not exercise if the price of IBM is less than or equal to $120.

In the case of this combination option, we may exercise this option (in some situations) even if the price of IBM is less than $120. Why is this so? We will exercise the first option at a loss if we believe the present value of the second option is greater than the loss we incur at the first option. To make this intuitively clear, imagine that 1 year from now the price of IBM is $115. When we come to exercising the first option, we know that we will immediately lose $5. Assume that Microsoft's price then is $32, and based on its characteristics our expectation of Microsoft's price 1 year after the first exercise (2 years from now) is $35. Meanwhile, our expectation of IBM 1 year later (based on its current price of $115 and volatility) is $120. This means that our expectation of the gain from the second option is 4 × 35 − 120 = $20. So, the loss of $5 now is far better compensated from the present value of the expected gain from the second option of approximately $20. We will exercise the first option and decide to gamble 1 year later for the second one. Note that we can only get an "expectation" of prices for IBM and MSFT at a later time, so nothing is guaranteed. It is possible that MSFT moves down from its price of $32 by the second year and IBM moves up. So, by the time the second option exercise time arrives, the gain from it is lower than what we anticipated, or even worse, the second option matures worthless.

Let us completely understand the mechanics of this decision 1 year from now as this forms the basis of many of the complex decision options problems

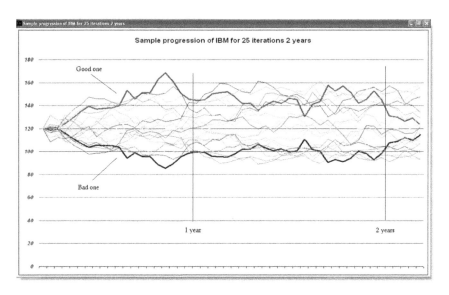

FIGURE 7.6
Stochastic simulation of IBM stock price.

described in this book. First, visualize the risk-neutral price paths for IBM in the 2-year horizon. Both the 1-year time point and the 2-year time point are shown in Figure 7.6.

Consider the scenario marked "Good one" in Figure 7.6. In this case the IBM stock price rises rapidly in the first year. Although it gets a bit tepid toward the end of first year, it still looks very good at $145/share by the time the first option is mature (ready to be exercised). So, it has risen a whopping $25 in the first year, nearly a 20% gain from the original $120 stock price. IBM shareholders are certainly happy. For the holder of the option combination also, this is an easy decision. The holder will exercise the first option, can immediately sell the share of IBM at $145, and can pocket the $25 gain. The holder maintains the option to make more money in the next year by exchanging a share of IBM for four shares of MSFT. When the second option matures in the money, the holder can always buy a share of IBM to exchange for the four shares of MSFT. So, the holder's overall gain can only be higher than the current $25 as the minimum the holder could make in the second year is zero as it is an option.

Now, consider the scenario "Bad one." IBM prices collapsed in the first year, and although they came back a bit toward the end of the year, they still lost $25 to a price of $95. This is nearly a 20% loss from the original price of $125. IBM shareholders are screaming at the management and considering replacing the board. For the holder of the combination option, this is a much more complex decision. If the holder does not exercise the first option, the holder forfeits the opportunity to exercise the second option. The first option is out of the money, so if the holder exercises (note that if it were a single option, the decision is trivial, and there is no exercise), he or she immediately loses $25. You may ask why the holder would even consider exercising the first one at a loss. The answer is that if the holder believes that the gain from the second option will be higher than the loss from the first, it may become optimal to exercise the first and take a loss. To assess the chances of making money from a combination option, the holder has to estimate the price of IBM and MSFT shares by the end of next year (2 years from the start of the problem). Also note that the exercise decision depends on the expectations of IBM's price 1 year from the first exercise as well. The exercise of the first option gives the holder a second option to exchange a share of IBM for four shares of MSFT. The holder does not have to own an IBM share for the exercise of the second one. If the second option is in the money, the holder can simply buy another share of IBM in the market and exchange that for four shares of MSFT. More likely, the writer of the option will settle with the holder for the net proceeds (4 × Price of MSFT – Price of IBM) without any open market transaction.

Consider the "Bad one" scenario from the risk-neutral price paths for IBM stock price again. At the 1-year mark, when the holder is considering whether to exercise option 1, he or she has observed the price of IBM to be a pathetic $95/share. Taking this and the stochastic characteristics of the IBM stock price, the holder can now estimate what the price is likely be by end of year 2 (Figure 7.7).

Note that such "estimation" is never precise (if it were, I would be writing this while sitting on a beach in Bermuda), but it is a better estimate than what the holder would have arrived at 1 year ago when he or she bought the

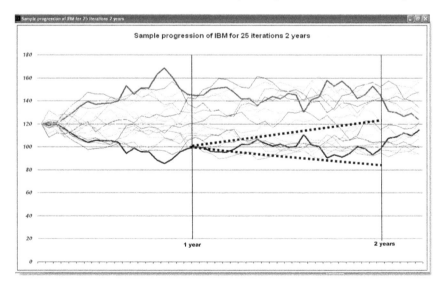

FIGURE 7.7
Forecasting IBM stock price range after observing first-year price.

combination option. The improved estimate will be within the cone shown in Figure 7.7. This cone of possibilities is much tighter than what it would have been at the start of the problem. Remember that the outcomes for IBM's price will be a wider band the longer into the future we need to project. At the inception, there were 2 years to the maturity point of the second option. At the exercise point of the first option, 1 year has already passed, and only one more year remains. The holder knows what the price is at the 1-year point, and the projection for IBM price is within a tighter cone than the one that existed at inception. The projection can be done by initiating another stochastic simulation with an initial price of $95 (observed price at year 1) and a volatility of 20%. This simulation has only 1 year to run, and we can estimate the probability of different outcomes of prices the next year.

However, a better estimate of the IBM price 1 year hence is not enough to make the exercise decision on the first option. The holder also has to have an estimate of the MSFT price as the second option is to exchange one share of IBM for four shares of MSFT. Figure 7.8 shows scenarios of risk-neutral price paths for MSFT.

Let's imagine that while IBM managers are having great difficulties in shoring up falling prices, MSFT is having a great year. The new operating system, in spite of its tendency to crash when humidity is high, turned out to be quite popular with customers. The price path is the one indicated by the arrow in Figure 7.8. It has risen to $50 from the starting $30, a mind-boggling 40% increase in the first year. MSFT is firing on all cylinders, it appears. At the 1-year period, the holder of the combination option can also witness this outcome. The holder can now estimate what MSFT prices

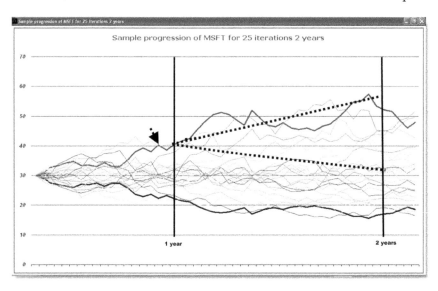

FIGURE 7.8
Stochastic simulation of MSFT stock price.

could be by the time he or she reaches the exercise decision for the second option. Just as in the case of IBM, the "cone of uncertainty" is smaller because the holder can now observe both the 1-year price as well as the fact that only one more year is left for the exercise of the second option. Again, this estimation can be done by running a stochastic simulation for MSFT price with an initial price of $50 (1-year price) and a volatility of 25% for the remaining 1 year.

Assume that, utilizing analytical techniques, the holder has "estimated" that the expected IBM price at the end of 2 years is $102, and the expected MSFT price at the end of 2 years is $55. These estimations are done at the maturity of the first option as he or she is trying to make a decision whether to exercise the first option (and take a loss) to keep the ability to exercise the second option alive. Since the second option is to exchange one share of IBM for four shares of MSFT, the holder can now estimate the gain from the second option at the time of the first exercise. The expected gain is $4 \times 55 - 102 = \$118$. The present value of this second-year expected gain is $118/1.02 = \$115.50$ as the risk-free rate is 2%. This far exceeds the loss the holder will take from the exercise of the first option (−$25), so the holder will gladly exercise the first option, cross his or her fingers, and root for IBM management to continue to have problems and MSFT to have a great year. After the exercise of the first option, the holder can sell IBM. When the second option is mature, the holder can simply settle it as a net transaction with the writer of the option.

Now let us examine what actually happened. As luck would have it, IBM managers got their act together, the new computer with a titanium frame is a good seller to corporate customers, and the stock has risen to $115 (from the 1-year price of $95). MSFT had problems with the new operating system (as it began to crash every Monday morning), and the stock took a slight tumble to $45 by the time the second option matured. The value of the second option at maturity is $4 \times 45 - 115 = \$65$. It exceeds what the holder lost from the first exercise, so he or she made money on the combination option, albeit much less than what the holder estimated while exercising the first option. You can easily imagine situations when the holder will also lose money by exercising the first one and the second option matures out of the money, thus netting a loss.

Visualize the risk-neutral payoff from this combination option. As mentioned, this is only a conceptual view (risk-neutral world), and actual payoffs will be different in the real world as both MSFT and IBM have a price process that has a drift commensurate with the systematic risk in their equity (which includes both operating risk and financial leverage).

The probability and cumulative probability of the risk-neutral payoff of the combination option being analyzed are shown in Figure 7.9 and Figure 7.10, respectively. Note that we value this option today at approximately $25.

A number of properties of this option and risk-neutral payoffs are worth noting.

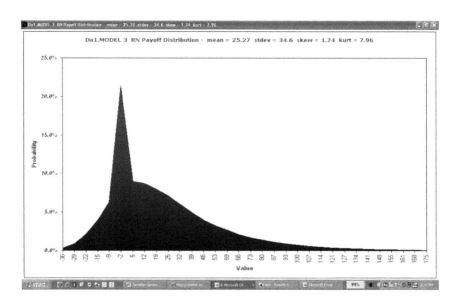

FIGURE 7.9
Risk-neutral payoff of the combination option.

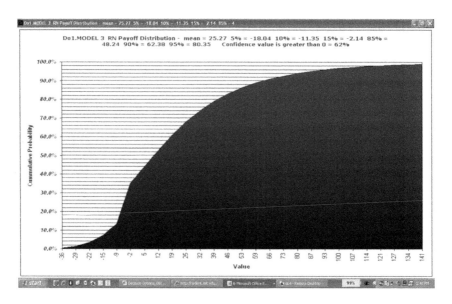

FIGURE 7.10
Cumulative probability of risk-neutral payoff of the combination option.

1. The option is highly valuable: priced at $25, although it is currently at the money for both options. Intuition may tell us that an option with an uncertain strike price (second option has a strike price equal to the price of IBM 2 years from now) is more valuable than one with a constant strike price. Because of the asymmetry in payoff (we can choose not to exercise if the strike price is greater than the asset price), higher volatility would imply a more valuable option. An option with volatility in both asset and strike prices will thus be more valuable.

2. The risk-neutral payoff shows negative values. This may be a bit confusing as we have so far established that options do not have negative values. But in this case, negative values do occur for the combination option because of the "gamble" the option holder has to make 1 year from now. Remember that one year from now, the holder has to make an educated guess regarding what the second option might be worth, combine that guess with the first one, and make a "go" or "no-go" decision. Sometimes it works out, and sometimes it does not. So, the negative values are those instances (in simulations) when the exercise decision taken at option 1 was the "wrong decision" and results in a loss of value. The risk-neutral payoff does not take into account the "price" the holder may have paid in "buying" this option combination at inception. So, the actual loss will be the loss from the options added to the price the holder paid for buying it.

3. There is a pronounced 20% probability peak at value 0. This corresponds to the no-go decision the holder may make at the exercise of the first option. This represents those instances when he or she decided not to exercise the first option at year 1 (after observing IBM and MSFT prices then and making an educated guess regarding what might be optimal). Note that the price of the option is "lost" in any case, and 0 value means that the holder does not lose any more. In this case, the holder decides not to exercise the first option and automatically walks away from the second one as well.

4. Although the option combination is very valuable (worth $25 today), there is only about a 60% probability that it will result in a positive payoff and only about ⅓ chance that the value will be above $25 and the holder will break even. As we have seen, this is a unique characteristic of options: there are small probabilities that large payoffs happen, but the most likely outcome is that the option expires worthless. Of course, this is a function of the strike price. If the strike price is a lot lower than the current price, the option will be highly in the money and will behave less like an option as the exercise is virtually guaranteed.

It is also possible to calculate "trigger price" combinations for both MSFT and IBM at the first decision point, which will make exercise of the first option optimal. In this rather trivial combination option case, exercise is optimal in all scenarios in which the IBM price is higher than the strike price. The holder of the combination option can just exercise the first option and pocket the difference between the price of IBM and the strike price. The holder is under no obligation at the end of second year to exercise the next one, and the least he or she could make from the second option is zero. However, if the price of IBM is lower than the first option strike price, it becomes more complex. Certain combinations of the first-year IBM and MSFT prices will make exercise optimal, and certain other combination of prices will make abandonment (walking away from both options) optimal.

Now, let us look at a problem that is not based on financial assets but on "real" assets. Assume that we are interested in buying a piece of land in the neighborhood. We do not, however, have the money (or the inclination) to buy it outright. What we would like to do is to buy an option to own the land at a later time. Let us make this a simple option. We would like to enter into a contract with the owner of the land to have the right (but not the obligation) to buy the land for $10 million 1 year from now. We can do market research and determine (at least approximately) the current value of the land. Assume this is $10 million. We also need volatility for land prices to calculate the value of this option in the contract. We can look at the value of this land and similar properties in the area over time. Volatility is a measure of how much prices move around over time. If we have historical prices of the property (or similar properties), we can simply take the standard deviation of $\ln(P_{t+1})/\ln(P_t)$ where P_{t+1} is the price at time $t + 1$ and P_t is the price at time t. Assume that we have done this, and the volatility is 10%. The risk-free rate (typically the yield on a 1-year T-bill) is 2%. Figure 7.11 shows the model.

We price this option at $0.5 million. This means that we will be willing to enter into a contract with the owner of the land that will give us a right to buy it 1 year from now at $10 million if the contract price we pay at inception is $0.5 million (or less). In the options parlance, the owner of the firm will be "writing" this option, and we will be buying it. For the owner, this is actually a way to create an income for this year while holding on to the asset for another year. Once the contract is signed, the owner will actually be hoping for a recession and a general fall in property prices in the area. This will mean that by next year, the value of the land will be less than $10 million, and we will just walk away from it (deciding not to exercise the option). In that case, the owner is better off by $0.5 million, the sum we paid to enter the contract. On the other hand, if the economy really takes off or there is more development in the area that enhances the value of the property to something more than $10 million, we will exercise the option. Note that we will only break even if the price of the land exceeded $10 million plus the price we paid to enter the contract. The $0.5 million that we calculated as the price of the contract is the efficient value. In negotiating with the owner of the land,

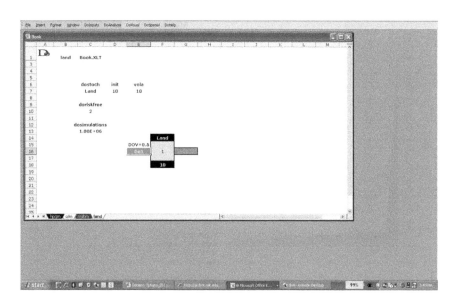

FIGURE 7.11
Valuation of the option on land.

we will like to negotiate a price less than $0.5 million. If we do pay $0.5 million to enter the contract, we are not expected to make any profit from it.

Figure 7.12 indicates the risk-neutral payoff of the contract. Note that the actual payoffs will be slightly different, and this representation is useful to assess the probabilities of certain payoffs. Many options purists may raise a number of objections about this example. The first objection will be that the underlying asset may not be replicable. Remember that we are relying on the risk-neutral pricing assumptions here to price the option. Risk-neutral pricing is valid only if a "replicating portfolio" can be created and if markets are complete and no arbitrage is possible. Market completeness in this context means that the investment does not expand the investment opportunity set (it is already part of it), and arbitrage is not possible when investors will quickly and costlessly move the price of the replicating portfolio to that of the asset itself.

To effect risk-neutral pricing, we need to be able to create a synthetic bundle that contains the underlying asset. In the case of a traded stock, this was easy. In the case of this real estate property, it is not as straightforward. The second objection is the assumption of GBM. We assumed that the markets are complete, and that the prices of land will follow GBM just like a traded stock. Although the assumptions that underlie risk-neutral pricing may appear daunting for many real assets, all of the said conditions can be met if a market can be conceived for the underlying asset.

We revisit this in number of places in this book and continue to provide further reinforcement to the idea of risk-neutral pricing and why it is perfectly

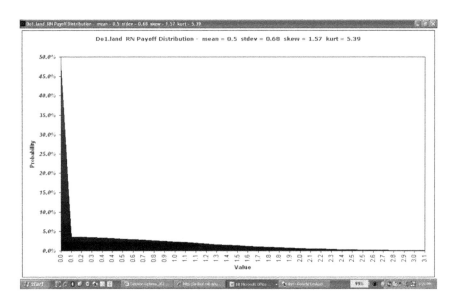

FIGURE 7.12
Risk-neutral payoff from the land contract option.

acceptable in all cases even if we cannot precisely create a synthetic bundle to match the payoff of the option.

We used GBM for convenience here, but if one suspects mean reversion in land prices, one can always find the rate of mean reversion and use the mean reverting stochastic process to represent the land. You may be wondering why land prices may show mean reversion. Land is a special kind of commodity—it is not completely divisible or substitutable. For land, location does matter. However, land prices are a function of real estate demand. Real estate demand includes the demand for residential and commercial building space. If demand is high, developers will buy land and start land development projects. This leads to a higher supply of residential and commercial building space, which will in turn drive prices down. As prices come down, properties become affordable for first-time buyers, and that in turn drives demand up. In addition, demand for properties is driven by mortgage interest rates, which in turn are tied to interest rate and the state of the economy. Demand typically increases as the economy heats up. In that case, the Federal Reserve, fearing an uptick in inflation, may move to increase borrowing rates. This will in turn increase mortgage rates, and this will dampen demand. So, there are multiple effects that may introduce mean reversion into land prices. As we have seen, that does not violate any principles of risk-neutral pricing. We are, however, assuming that markets are at least weak-form efficient. If the weak form of the efficient market hypothesis (EMH) does not hold, technical analysts would have amassed the world's wealth by simply exploiting "patterns" in asset prices. Since this has not yet happened, it appears that EMH is a safe and conceptually elegant assumption.

Let us further complicate the land contract. Assume that we are considering a contract on the same property with the following characteristics. In 1 year, we have an option to put in $2 million to have a right to buy the land 2 years from now. This $2 million is like a "deposit" for the owner. If we do pay the deposit 1 year from now, we keep the option to buy the land for $8 million 2 years from now. However, there is a zoning decision that is expected 18 months from now regarding this particular property. The property is currently zoned commercial, and that is status quo. However, there is a 10% chance that it will be rezoned and categorized as noncommercial. In this case, we forfeit the option and get back half of our initial deposit of $2 million. If the status quo zoning (commercial) is maintained, then we can exercise the option to buy the land for $8 million. All characteristics of the land price processes are as before (it is the same land). Figure 7.13 is a decision options model for this problem.

As you may have guessed, this problem has two options: one at the end of the first year (the decision to pay the deposit) and the other 2 years from now (the decision to buy the land). These are options because we have no obligation to do either. We do not have to make these decisions now; they can be made at a future time. They are dependent on land prices that we can observe. They also depend on a private risk, the chance of a zoning change. We have a probabilistic expectation of this event, and we get no information on how likely or unlikely this event is over time. We will only know after the event. The second decision to buy the land, 2 years from now, is a simple one. By then we will know if the zoning change has happened and

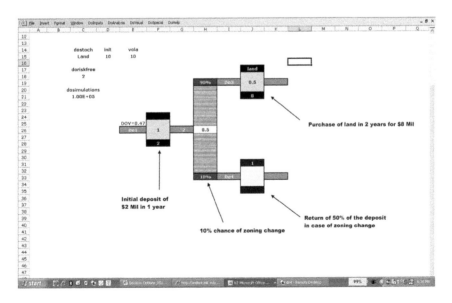

FIGURE 7.13
Decision options model for land contract.

the market price of the land. If the zoning change has happened, the option to buy the land does not exist any more, and we get $1 million back. In that case, the decision is made for us (a predetermined outcome). However, if the zoning change did not happen, we hold an option to buy the land for $8 million (assuming we made the deposit the prior year). To exercise the second option, we will simply compare the market price of the land (in 2 years) and $8 million we need to pay to acquire the land. If the market price is higher than $8 million, we will exercise the option and become the proud owners, and if the market price is lower than $8 million, we simply walk away and take solace in the fact that we lost only $2 million and perhaps we can write that off on a tax return. If we do buy the land, we can also immediately sell it at market price and pocket the difference (Market price − $8 million). Depending on our credit rating, we may not even have to put up the $8 million if the transaction is going to happen immediately.

The decision to invest a $2 million deposit 1 year from now, to keep the buy option alive, is a much more complex one. By the time the first year comes around and the option to make the deposit becomes mature, we have new information on land prices. We can observe the market price of the land then. This price is likely different from what it is now. However, we do not know what the prices will be when we are ready to buy it (1 year hence) or if the regulators' finicky policy making will result in a zoning change. As one can imagine, anticipating policy is a real gamble.

The decision to pay a deposit is similar to the financial instrument MSIBM we discussed. When the time arrives to make this decision, we can observe the market price of the land and, based on that, anticipate what the prices will be 1 year later. We then calculate the gain from the second option, make an appropriate correction for the private risk (the chance of zoning change), and then ask whether it is worthwhile to pay the deposit. We will only do so if the present value of our expectation of the future gain (adjusted for regulator whim) is higher than the $2 million we need to spend. Note that in spite of our best intuition and analysis, it is possible that we may be wrong, so this is an educated guess. We do have the benefit of observing the land prices 1 year from now before making the decision, and we have no reason to make it now. You may want to contrast this situation with what is typically done in traditional DCF.

Considering the uncertainty in land prices and future decision flexibility embedded in the options, this contract can be valued at $470,000 currently. This means that we are willing to pay "up to" $470,000 ($0.47 million) to enter this contract today. In other words, if we enter the contract today with no initial payment, we are expected to gain $470,000 from the entire deal.

Figure 7.14 is the risk-neutral payoff from the contract. The high peak of 45% centered on 0 implies that there is a very high likelihood that we will never pay the deposit in the first year and will call it quits. This may seem counterintuitive to some. Why would the contract be worth $470,000 today if the most likely outcome is walking away from the deal in 1 year? The answer

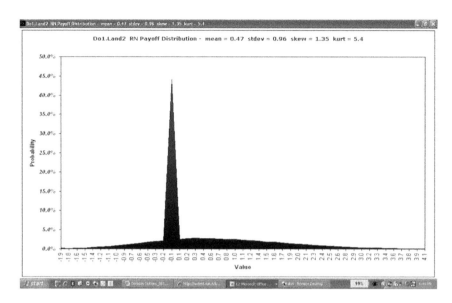

FIGURE 7.14
Risk-neutral payoff from land contract.

is that 1 year is a long time, and you do get a lot of information during that time to make a better decision. Paying the deposit is a complex decision that involves evaluating how land prices are doing and our chances of making a fair return at the end of the contract. The counterparty in this contract knows that the option to deposit at a later time as well as the option to buy 2 years later are valuable, and he or she will likely demand some compensation up front for such a contract. Our analysis shows that we may want to negotiate this up-front compensation, but if we pay anything more than $470,000, we will lose. We may want to negotiate our way up from zero and walk away from the contract if he or she demands more than $470,000 up front.

The risk-neutral payoff diagram in Figure 7.14 shows that in 45% of the cases, we decide that it is better just to walk away and not put up a deposit at the end of first year. But, in the other 55% of the cases, we do go forward. However, we may lose our deposit completely if the zoning did not change but in 2 years the land prices drop to less than $8 million, so the second option expires worthless. In some cases, we get half the deposit back (but still lose the other half) as regulators in their infinite wisdom turn the dials and make a zoning change. In some cases, we do make a profit from buying the land as the market price is higher than the $8 million purchase price (second option), but it may not be enough to compensate for the deposit of $2 million, so we still end up losing some money (but not all the $2 million). However, it is not all bleak. In some cases, we make a killing: no zoning change and land prices really take off. It is for these "killer" scenarios that we actually buy an option.

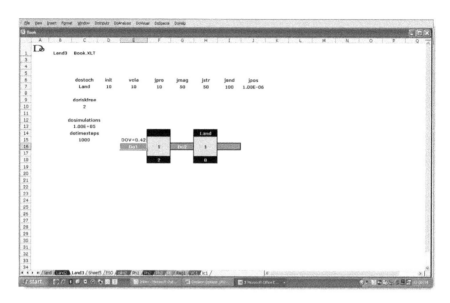

FIGURE 7.15
Land contract with jump in land price process.

Since the landowner already knows that we will only exercise if the market looks good and simply walk away from the deal otherwise, he or she will likely want an up-front fee to sign off on such a contract. The type of analysis conducted here does give the option holder a reserve price (a price beyond which he or she is likely to lose money) and a great advantage in negotiations.

To understand the modeling architecture, let us also consider another similar problem in which the two options are the same. Instead of a zoning change probability of 10% (and associated refund), we assume that at any time between the first and second year there is a 10% chance a zoning change may occur. If such a change occurs, the value of the land prices drop by 50%. We model this as a jump in the stochastic process. Figure 7.15 shows the model. As can be seen in the model, the contract is worth $0.42 million ($420,000), slightly less than the contract we discussed.

Given in Figure 7.16 are a few samples of the simulation of the land prices. Note that 2 of the 25 samples show catastrophic drops in prices between the first and second years. We modeled these jumps to occur any time after the first decision. In these two cases, soon after writing the check for the deposit, we may end up cursing our luck as the drop will almost ensure that the prices cannot climb back to a level that will make the purchase profitable. So, even though we may have done a good job anticipating prices based on the market, we still cannot win because of the arrival of the jump. Life is interesting but never fair.

Figure 7.17 shows the risk-neutral payoff from such a contract. As can be seen from the distribution, in some cases the $2 million deposit is lost, and

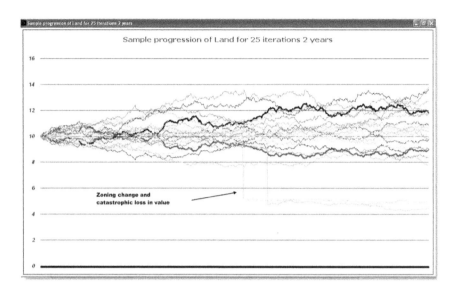

FIGURE 7.16
Stochastic simulation of land prices with jumps.

there is a high probability of walking away from it without paying a deposit when the first year comes around.

As you may have gathered, what is different in all these examples from traditional finance is that there is flexibility in future decisions. When we come to a future decision point, we have more information on the assets involved, and we have an option—the right but not an obligation—to do something. This uncertainty and evolving information coupled with decision flexibility makes evaluating such investments different from the rigid assumptions of DCF analysis based on the capital asset pricing model (CAPM). It may be worthwhile to revisit the assumptions underlying the CAPM-based DCF. When we have a set of future cash flows as in the example of the land contract problem, we assume the following:

1. All decisions affecting future cash flows are made now (no flexibility in future decisions).

2. If we are doing traditional DCF, we also assume there is no uncertainty. So, we take the average expectations and use point estimates of cash flows.

3. We also have to find a discount rate. This is where significant confusion still exists for practitioners. Remember that the discount rate is supposed to reflect the systematic risk. In the case of the land contract, somehow we need to find the correlations of expected return in the contract against a "market portfolio." We have multiple problems here. The market portfolio is not really observable (perhaps we can

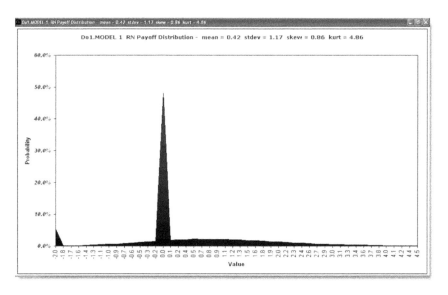

FIGURE 7.17
Risk-neutral payoff from land contract with jumps in land prices.

use some proxy such as the world index of stocks, bonds, and real estate investment trusts). To determine the discount rate, we then have to find the correlation of the asset's return against the market portfolio's return. Typically, a proxy that can be observed in the market is used. It is a complex exercise, and in many cases a proxy cannot be found. In case a marketed proxy can be found, one also has to be careful in calculating the asset β. To avoid all these complications, analysts tend to use the cost of capital of the firm (in this case, it will be our cost of capital, i.e., the cost of borrowing money for us) as the discount rate. Some throw in the towel and use 10% as it has a nice ring to it.

4. Some approach these types of problems by calculating an internal rate of return (IRR). The IRR is a metric that sometimes incorporates private and market risks in a single number (if cash flows are not adjusted for private risk before the IRR calculation). The process of comparing IRR to cost of capital or other thresholds to make a decision suffers from the same issues mentioned.

In any case, the first two assumptions—lack of uncertainty and lack of flexibility in future decisions—are clearly not the case, so DCF, NPV, and IRR are unlikely to be useful in making a decision.

8

Employee Stock Options as Decision Options

Many companies have been awarding employee stock options (ESOs) to managers and key employees for many years. You may be familiar with the back-dating of stock options scandal that emerged recently in some technology companies. Some companies may have engaged in the fraudulent practice of changing the characteristics of the employee options (time to expiry and strike price) when their stock prices fell and the options awarded to employees matured worthless. ESOs are similar to financial options with one unique difference: they are not traded and are typically exercised prematurely by the employees. As you know, it is never optimal to exercise an option prematurely if you can sell it. In the case of ESOs, holders may exercise them prematurely to diversify their portfolios or to create income. For many owners of the ESOs, a large percentage of their revenue and estate are tied to the prospects of the company for which they work, and diversification may become a strategic need. You may remember that many employees of Enron, who invested most of their 401(k) pension plan assets in Enron stock and owned ESOs on Enron stock, had a difficult time when Enron went bankrupt. There is no better wisdom than the importance of diversification in all areas of finance.

As one can imagine, these options are valuable; thus, companies that issue them are "paying" something to their employees although there is no cash payment. Since there is no cash payment, companies do not "expense" options in their financial statements and basically keep them off their financial reporting. Only when options are exercised do they show up as having an impact on the company's capital.

The following is an excerpt from the Financial Accounting Standards Board (FASB):

> On March 31, 2004, the Financial Accounting Standards Board (FASB) issued a proposed Statement, Share-Based Payment, that addresses the accounting for share-based payment transactions in which an enterprise receives employee services in exchange for (a) equity instruments of the enterprise or (b) liabilities that are based on the fair value of the enterprise's equity instruments or that may be settled by the issuance of such equity instruments. The proposed Statement would eliminate the ability to account for share-based compensation transactions using APB Opinion No. 25, Accounting for Stock Issued to Employees, and generally would require instead that such transactions be accounted for using a fair-value-based method.

The FASB recently made an accounting change, now requiring companies to expense stock options based on the fair value of these options. What is interesting here is the question of what may be fair value for these options. It is clear that since these options are not freely traded and are typically exercised prematurely (and thus suboptimally), they do not have the "full value" as one may calculate using traditional option pricing models such as the Black-Scholes equation. However, the value is clearly not zero; otherwise, employees will not happily accept them or companies issue them to "retain" and "motivate" their best employees.

How can we calculate the value of ESOs? One way to do this is to simulate the price process of the stock and artificially impose early exercise rules to create the value of the ESO. Figure 8.1 shows a software tool that allows companies to value their ESO awards based on the exercise patterns of their employees.

An ESO has the same fundamental attributes of a call option, such as the strike price and the time to expiry. However, it has some added features. For example, it cannot be exercised before it is vested and can be exercised any time after the vesting period is over.

In Figure 8.1, the interest rate, dividend, and volatility can be stochastic. This means that instead of a constant risk-free rate, dividend rate, and volatility, we can actually provide a "term structure" in risk-free rate and dividend rate. They vary with time, and we have expectations around how they are going to change in the future. For example, the term structure in risk-free rate can be observed in the marketplace by using the yield on risk-free instruments such as T-notes and T-bills with differing maturities. Similarly, the company may have a dividend policy in place with increasing (or decreasing) dividends in the future. We can also derive a time-varying volatility by using historic volatility or implied volatilities from market-traded derivative instruments. Implied volatility can be calculated from the prices of market-traded options of differing maturities on the company's stock. If such market-traded instruments exist, it is always better to derive an implied volatility and use it in pricing rather than falling back on historical volatility as markets always look forward. As we all know, driving a car using images in the rear view mirror is always risky.

For ESOs, there are additional complications related to premature exercise. Employees may exercise them prior to expiry, thus driving down the value. As may be clear, premature exercise of options always decreases the value of the option (except in certain special cases when dividends occur close to expiry). The company may have historical records of ESO exercise patterns from the past. If the employee pool does not change dramatically over time, the company can make an argument that the past employee exercise patterns will continue into the future. There are two types of premature exercise patterns seen in ESOs. First, employees tend to exercise when the stock price goes up a certain percentage. This is akin to timing the market by equity investors. The more the stock price goes up, the more likely that the employees will exercise. Second, the employees tend to exercise after a

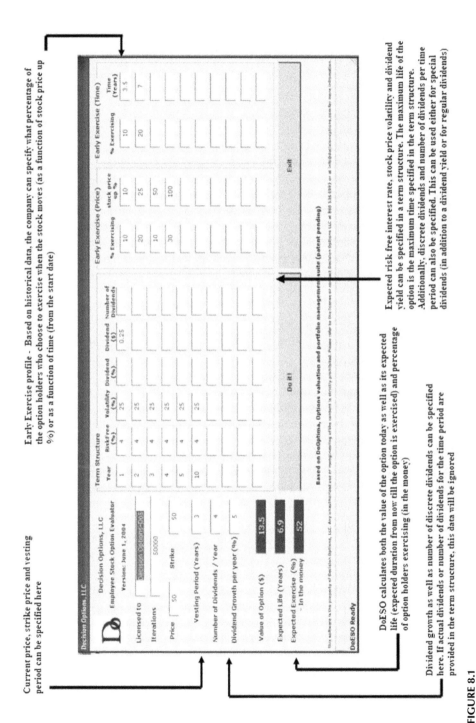

FIGURE 8.1
Employee stock options valuation tool.

certain time from the award of options either to create income or to diversify the portfolio. The "older" the options are, the higher the chance of exercise. By using historical data, a company can create an option exercise curve as a function of stock price growth (as a percentage of strike price or award price) and as a function of time (from the moment of award). As mentioned, estimates based on history are never perfect. This is especially true if the demographics of the employees (age, gender, and location) have changed over time. In such situations, the company may be better off estimating premature exercise based on broader industry data on ESOs than using its own historical records. In any case, the fact remains that employees will exercise prematurely.

To further complicate ESO valuation, the company may impose certain blackout periods during which time no exercise is permitted. There may be quarterly or yearly blackout windows as well as *ad hoc* blackout periods when there is significant news pending. These are constraints that need to be imposed on the ESO that may further drive the value down from a pure market-traded option. Constraints on exercising can have a significant impact on value because the ESO cannot be sold in the market.

A company will have pools of ESOs with varying maturities (as given to employees over a period of time). Each of these pools may share certain common characteristics such as time to expiry and strike price. To carry ESOs on its balance sheet, a company may first have to categorize all the unexercised ESOs in its book into pools that share common characteristics. Then, it has to value each pool and add them to determine total liability. A change in this liability may also be an expense (or unusual gain) every year. To expense the award of ESOs in any particular year, a company has to value what was awarded that year as well. So, to accurately reflect the effect of ESOs in its financial statements, a company has to value both its stock of unexercised ESOs and the new batches of ESOs being written. For many technology companies, ESOs may represent a significant "off-balance-sheet" liability as well as a big expense that will affect reported earnings per share (EPS). ESOs are transfers of shareholder value to employees.

To value an ESO, we first simulate the stock price path using stochastic simulation, assuming that the stock price follows geometric Brownian motion (GBM). In this simulation, risk-free rate, dividends (either continuous or discrete), as well as volatility are allowed to vary as a function of time. As shown in Figure 8.1, the ESO calculator, you can specify the risk-free rate, dividend rate, and volatility for every year in the future until the expiry of the ESO being valued. It is also possible to consider discrete dividends (as is typical). A company can specify its dividend policy for every year into the future until the expiry of the ESO. In the stochastic simulation of the price process of the stock of the company (on which ESOs are awarded), we use the appropriate drift (risk-free rate – dividends) at every time period as well as the appropriate volatility at every time period.

In the GBM equation we considered, we now have r, d, and σ as a function of time:

$$\ln(S_t) = \ln(S_{t-1}) + (r_t - d_t - \sigma_t^2)t + \sigma_t N(0,1)\sqrt{t}$$

The mechanics of simulation remain the same. Once we have the risk-neutral price paths of the stock price, we can then impose the early exercise patterns of the employees. Since we use the percentage change in stock price to model early exercise based on price, the exercise windows will remain the same whether the price paths are simulated using risk-adjusted discount rate or risk-free rate. Similarly, we can impose the early exercise pattern based on time. For every simulation, in every time interval from the award to expiry, we can calculate the price (using the term structure in risk-free dividend and volatility) and then exercise options according to the early exercise patterns. So, we get a value of the option in every simulation considering all the information provided. By doing this simulation thousands of times, we can calculate the expected value of the ESO.

In the example, the stock price at the time of award as well as the strike price is $50. The vesting period is 3 years; this means that the option cannot be exercised in the first 3 years. This constraint is applied by companies to ensure that the employees who have ESOs do not immediately exercise them and quit, defeating the reason for providing ESOs to motivate employees. The current policy of the company is four quarterly dividends of $0.25 each, so this stock currently has a 2% dividend yield. The management of the company believes that this dividend will grow at 5% per year as has been the norm in recent years. Based on its analysis, the company has decided to use a constant risk-free rate of 4% and a volatility of 25%. The ESO has a 10-year expiry period.

By doing a historical analysis of its employees' ESO exercise patterns, the company has created early exercise patterns. First, it looked at when employees exercise as a function of stock price. The following pattern was found:

% of ESO holders exercising	% of stock move up from award
10	10
20	25
10	50
30	100

This means that 10% of the employees exercised when the stock moves up 10%. So, if the stock moves to $55 from $50, the company expects 10% of the employees who received this award to exercise. Another 20% exercised when the stock price increased by 25%. So, a total of 30% of the employees would have exercised once the stock moved up by 25%. An additional 10% exited when the stock price moved up by 50%. When the stock doubles (up by 100%), a total of 70% would have exercised. The remaining 30% did not exercise until maturity and held on to the "full value" of the option.

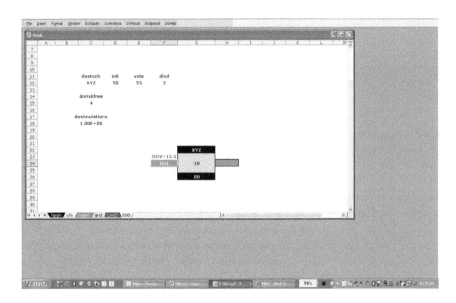

FIGURE 8.2
Full value of an ESO without constraints.

Similarly, the company created an exercise pattern as a function of time from award:

% of ESO holders exercising	Time (years) from award
10	3.5
20	7

The company found that 10% of the employees exercised (regardless of price) within the first 3.5 years. Note that the ESO had a vesting period of 3 years. So, this means that 10% of the employees exited very soon after the vesting period. Within the first 7 years 30% (20% added to the 10% who exited within 3.5 years) of the employees exercised. From the exercise pattern based on stock price, we have found that about 30% exercise when the stock has moved up 25%. So in combination, it appears that the stock moves up about 25% in 7 years.

Let us first calculate the full value of the ESO assuming no constraints and premature exercise. The vesting period of 3 years has no impact as the optimal exercise policy of the option is at maturity. The model in Figure 8.2 shows that the full value of this ESO (if held to maturity) is $15.5. A volatile stock combined with a long period (10 years) to maturity gives this option a huge value. In some cases, the employees may not fully appreciate the value given to them.

Please note the following about the inputs. First, we have used a continuous dividend yield of 2% instead of the company's discrete dividend policy

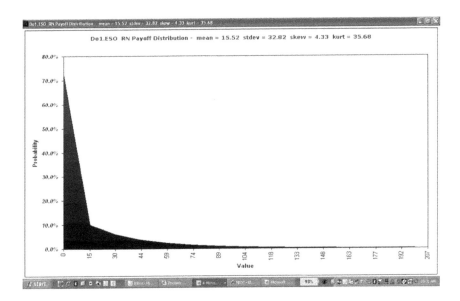

FIGURE 8.3
Risk-neutral payoff from an ESO if held to maturity.

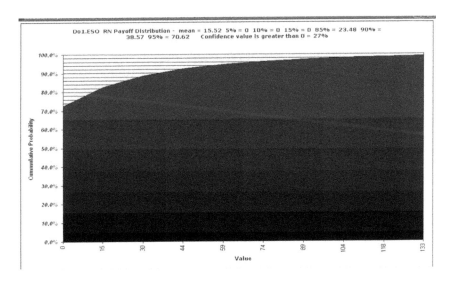

FIGURE 8.4
Cumulative probability of risk-neutral payoff from an ESO if held to maturity.

(as used in the ESO valuation). Second, we assume that the 5% increase in the discrete dividend policy the company has set forth is to maintain its dividend yield in the range of 2%. Note that dividends have to increase over time to keep the dividend yield as the company's stock price will also be increasing.

Also, given in Figure 8.3 is the probability of risk-neutral payoff from the option (if held to maturity). We find that nearly 73% of the time the option expires out of the money if held to maturity. It may be counterintuitive to you why the optimal exercise policy still is at maturity even with such a high chance of expiring out of money. The reason is the tail of the risk-neutral payoff. In the remaining 27% of the time, the option could be highly valuable.

We find that if held to maturity, there is only a 27% chance of making money on this option, and the expected value of the option is $15.50. The company that awarded this option may find it hard to "expense" the full $15.50 in its income statement as it knows a large percentage of its employees will exit much earlier than the 10 years given to maturity. Figure 8.4 indicates the cumulative probability of the risk-neutral payoff that shows that the chance of making money on the option is only 27%.

From the results on the ESO calculator, we find that by considering the early exercise patterns of its employees, discrete dividends, growth in dividends, and vesting period, the actual value of the ESO is $13.50. This is 13% less than the full value we calculated. We also find that expected holding period on the ESO is 6.9 years, and that there is a 51% chance that the option will be exercised in the money. Contrast this with the full value calculation in which the holding period was always 10 years (maturity), and only 27% exercise in the money. So, by exercising earlier (about 3 years earlier on average), nearly twice the employees (51% instead of 27%) made some money (more than 0), but in aggregate they made 13% less ($13.50 instead of $15.50). Note that we have excluded any information advantages for the employees in "timing" the exercise of their options. Also, unlike regular options (which can be sold), employees can only "exercise" to exit the position. So, exit is synonymous with exercise for the ESO, and that may justify the premature exercise behavior of some employees.

As some are aware, there has been a recent scandal involving many high-tech companies that "backdated" their ESOs. In this case, the backdating may work as follows. Suppose the company issued the ESO on January 1, 2003, at $50 strike when the stock price was $50. Five years later, on January 1, 2008, it finds that the stock has crashed to $25. Its key employees now know that their options are reasonably worthless as they have only 5 more years to maturity. Changing the parameters on the ESO after observing what the price has done is a bit like asking for your money back after losing a hand at the blackjack table. While obviously illegal, some companies have changed either the strike price or the time to expiry on the ESO. For example, they can "change" the strike price on this ESO on January 1, 2008, to $25, making it instantly highly valuable, or they can provide more time for the option to expiry, providing a higher probability that the stock will move back over the strike price. Both of these practices are illegal, and akin to taking money out of the shareholder's pockets and giving it to the employees in broad daylight. However, the illegality did not seem to deter the management of certain companies.

We started the ESO discussion to ascertain how much the company should expense when it awards an ESO. In the example we discussed, the "diminished value" of the ESO due to early exercise patterns of the company's employees is $13.50. So, the company should expense $13.50 for every such ESO given out. This is the true value of the ESO, and the award is not dissimilar to writing a check to the employee for $13.50 for each ESO. This is real compensation for the employee and a loss of value for the shareholders. By not recognizing the award in its financials, the company may give a false impression of "income" from its operations as it will "undercount" compensation expenses. For many technology companies and start-up companies, stock options are primary components of compensation for their employees. For technology companies, this is one way to find and retain highly motivated employees. For start-up companies, this is a method to save cash expenses up front as they have to invest in research and development before creating revenues. In either case, such options have to be valued and expensed in the financials of these companies for a truer picture of their financial position.

The market may have been correctly assessing the value of ESOs even before they showed up in the financials (as needed by the recent FASB accounting change). Companies afraid of making the accounting change should take comfort in the fact that the market already knows how to value ESOs, and the award details were always in the financials. It is unlikely that an accounting change will substantially change the value of the company to shareholders or the stock price. The value of the company is not based on backward-looking financial statements but rather forward-looking expectations of cash flows based on the company's strategy. Thus, the time spent dressing up the financial statements for cosmetic quarterly and annual EPS is a waste. It will be a lot better if companies publish all information in a consistent way and let the market figure out what the price of stock should be. In this context, any deterministic EPS forecasting and numerous subsequent adjustments to it are activities that add no shareholder value. Avoiding such activities could provide executives with more time to think about strategies that have positive impacts on the company's value.

9

Case Studies in Life Sciences

The U.S. life sciences industry—a category that includes pharmaceuticals, biotechnology, and medical equipment companies—invests over $100 billion a year on research and development programs that aim to discover and develop new therapies to prevent, diagnose, cure, or alleviate diseases affecting humans and animals. These expenses are spread over a long time, often decades, from idea generation stage to marketed product. And, with only 1 of over 100 ideas turning into a commercially viable product, the risks are enormous.

The life sciences industry (including pharmaceuticals, biotechnology, and diagnostic equipment) generated total revenues of over $800 billion in 2007. The pharmaceuticals segment was the industry group's largest in 2007, generating total revenues of over $600 billion, equivalent to 75% of the industry group's revenue. The performance of the industry group has been forecasted to decelerate, with an anticipated CAGR (cumulative annual growth rate) of 8% for the 5-year period 2007–2012. The group is expected to generate revenues exceeding $1.2 trillion by the end of 2012.

Despite the robust revenue numbers and robust growth forecasts, the pharmaceutical industry has been in a perfect storm for nearly a decade. The factors that have brought it from a business with one of the highest returns on investment (ROI) to one with the lowest are both external and internal. However, all of these factors have been exacerbated by a management style that is still largely operating as if the environment had not changed. It is imperative that the industry move into better ways to select and design projects and systematically manage a portfolio of projects to enhance shareholder value. More important, with the world's population aging and the risk of bioterror increasing, life sciences products take on a more important role in the survival of our own society. To move back to their former glory, life sciences companies have to break away from the shackles of conventional management and traditional ideas perpetuated by consulting companies and advisory firms. Mergers and long research-and-development (R&D) and product cycles in this industry have led to a lack of accountability and market test for management decisions.

The life sciences industry not only has to manage significant uncertainty (technical, market-based, regulatory, demographic) but also has several degrees of management flexibility to manage through it. The decline in the value of the pharmaceutical sector has now created an acute pressure that needs a response. Clearly, the traditional techniques such as cost cutting through reduction in the size of R&D spending will not fix the problem.

R&D programs generally have a macro-objective, such as finding a cure for Alzheimer's disease using a specific mechanism and technique. To meet the objective, R&D programs target to develop a number of different "candidates" that may fit a general profile. Each candidate has expectations in terms of the end products, delivery mechanisms, pricing, and other factors. Each candidate also has a project plan with various investment options and schedules, and some plans contemplate expansion of the product into other disease areas, population groups, and delivery mechanisms (dosage forms). To reach their goals, the project plans must coordinate a number of different "specialty" areas or disciplines, including clinical, toxicology, chemistry, biostatistics, substance manufacturing, and pharmacy. Adding further to the complexity, the development plan for each candidate may also include many partners outside the company (research collaborations, universities, contract research organizations (CROs), contract manufacturers, governments, regulators, insurance companies, physicians, and patients) who may provide resources and services specific to the program. It is indeed a massive undertaking, with some development plans surpassing the gross domestic product (GDP) of small countries in overall required investments. The complexity of these development plans creates significant uncertainty about outcomes, but along with such uncertainty comes considerable opportunity to build managerial flexibility into the design and execution of such programs.

The value of life sciences companies is primarily driven by their internal R&D. As previously seen in many other industries, neither acquisitions nor *ad hoc* cost reductions increase their value. In acquisitions or selective licensing of products from other companies, the price paid generally exceeds value gained and thus can only result in shareholder value loss. Cost reductions, if indiscriminate and *ad hoc* (as they typically are), result in collateral damage in value-producing entities. In this people- and information-intensive industry, cost cutting generally results in loss of skilled people and lack of overall flexibility for future growth. The market, recognizing this, generally welcomes R&D cost cutting by selling the stock of the company.

Economic analysis and empirical observations show that over 50% of the market value of a pharmaceutical company is related to growth options from internal R&D. In other words, a traditional financial analysis of the company can explain only 50% of the market value. This means that the market fully appreciates the optionality in internal R&D in pricing the company. Contrary to popular belief that Wall Street is obsessed with short-term earnings per share (EPS), evidence shows that markets are generally forward looking in valuing a company. The efforts spent by the finance organizations in precisely predicting and "meeting" EPS is an activity that destroys shareholder value in the long run (at least by the wasted effort expended).

Life sciences companies have long recognized that their primary assets are the R&D programs in the pipeline. Although most managers have an intuitive understanding of the value of R&D programs, the quantification of that value has been a challenge for a variety of reasons:

1. There are conceptual misunderstandings about what is meant by value.
2. Traditional techniques such as discounted cash flow (DCF) and decision tree analysis (DTA) do not capture the uncertainty and flexibility inherent in R&D programs.
3. The availability of tools to conduct valuation based on more generalized and appropriate frameworks has, thus far, been limited.
4. Existing managerial incentives within departments may prevent needed changes in decision processes.

The industry practice in external transactions such as licensing and contract manufacturing has been to use rules of thumb developed over the years. At the same time, however, "internal transactions" such as project selection, prioritization, and resource allocation are evaluated primarily using traditional techniques such as DCF and decision trees. Decision makers understand that current practices are limited and potentially misleading, and they attempt to compensate for the limitations of these techniques by making qualitative adjustments based typically on the technical aspects of the program.

The following are fairly reliable indicators that the concept of value is not understood or not applied in a consistent way in valuing R&D:

1. *Reliance on rules of thumb or proxy-based licensing and contract deals with external partners.* This is similar, in many respects, to the valuation practices used in the venture capital industry. Factors such as the size of the overall market, the reputation of management, and any available technical data tend to be the main drivers of transactions. Initial bid prices, which typically take the form of milestones and royalties, are based on those in previous transactions or, more likely, a single successful transaction in the past. There is often some negotiation around such rules, but almost always within limits established by sacred cows (e.g., "absolutely no milestone payments at filing"). Discussions and debates focus on technical details (e.g., "the rat study is very promising, and that is the primary basis of this deal"). How all this translates into shareholder value added is almost never mentioned. Moreover, the split of value between the licensee and licensor created by the deal structure is also generally not subject to much analysis. The tacit assumption underlying such practices is that if the deal is structured according to the rules of thumb, it is a good deal for the company.

2. *Prioritization of programs (and entities) using rankings and multiple evaluation criteria.* R&D programs are complex and provide significant flexibility for multidimensional rankings (on criteria such as safety, efficacy, manufacturability, differentiation, and the cost of

raw materials). Traditional decision trees have been used widely in pharmaceutical companies, and they can be used to calculate the same net present values (NPVs) produced by DCF analysis. But, such NPVs are generally viewed by decision makers as only one of several criteria—one that may be too narrowly "financial" to capture the "spirit" of the program under consideration. Moreover, the use of traditional decision trees has led to some confusion in the marketplace as some practitioners have mistakenly labeled it "options analysis" (to denote the branches in the tree). As most first-year business school students know, traditional decision trees are simply pictorial representations of the mechanics of the DCF analysis, nothing more. That is, they do not consider all uncertainties present or incorporate the impact of management flexibility that is inherent in the decision process.

3. *Resourcing (budgeting) decisions are segmented by departments, products, and specializations and are generally based on last year's budgets.* Since there is no common currency to compare investment choices across departments, products, and specializations, resourcing decisions (budgets) are typically done in a segmented fashion. Resources are allocated into buckets, typically according to a formula based on overall sales, last year's budget, and growth rates. Once a departmental (or product) budget is set, managers further divide that amount based on local formulas. Such allocations in turn typically depend on last year's budget or on managers' negotiating skills. It is not unusual to find strong correlations between departmental budgets and the seniority and education of the manager.

It is intuitively clear to decision makers that every product or investment opportunity has an intrinsic value to the company. It is also clear that investment opportunities may present various paths forward, and each path (or design) may have different values. If a method could be established to systematically value every investment decision (including alternative designs), one could create a common currency for use in selecting, comparing, prioritizing, designing, buying, and selling investment opportunities. If the method is applicable across all investment choices, the common currency of value can be the only decision variable regardless of the nature, location, time horizon, size, and so on of the investment choices that are available. This is because value, if calculated using an economically consistent method, would capture the information related to all parameters and the uncertainties in the estimation of those parameters. Moreover, the valuation method should be roughly consistent with the intuition and thought process that experienced decision makers go through when they select the best opportunities.

To put this in the right context, consider a pharmaceutical company with the following types of investment choices in R&D:

1. A full development candidate entering phase III (large-scale clinical studies undertaken after proof of concept has been established)
2. An early development candidate that has just filed an IND (Investigational New Drug) application and is ready to enter the clinic (human trials to assess safety)
3. An IT (information technology) infrastructure improvement project that is expected to enhance productivity in record keeping
4. An expansion of a pilot plant that requires significant capital expense
5. Hiring of new personnel with specific expertise in oncology
6. A licensing opportunity with a biotech company on a candidate in an area in which the company has its own program

Suppose also that the company has a hard resource constraint (a limited budget), either imposed by senior management on R&D or set by market forces on the level of R&D that appears optimal at the current time. The question is how such resource constraints should affect investment decisions inside R&D. How does the company decide which investment opportunities to select and prioritize? How and when should the company execute the projects it decides to undertake? In a traditionally managed company, investment opportunities will be selected, prioritized, and funded by different departments in a largely uncoordinated process, and as mentioned, the budgets for those departments are likely to be determined mainly by last year's budgets. Such segmentation introduces the possibility that the best opportunities, if located within the wrong department, may be underfunded or passed over completely for "lack of budget." To make matters worse, finance departments may make tactical adjustments to the departmental budgets to improve quarterly financial statements (apparently believing that investors focus mainly on the next quarter's earnings). As such tactical allocations (cutbacks or increases) flow through departments, they further affect the optimality of the investments undertaken by department managers.

In sum, existing practices and incentives are likely to reward management more for the ability to negotiate budgets than to add value for shareholders, resulting in overinvestment in some projects and underinvestment in others. One requirement for changing such practices and incentives is to devise a better measure of the value added by R&D projects. But, this, of course, is far easier said than done since the payoffs from such investments tend to be realized years after the initial decisions are made, and managerial incentives tend to be much shorter in time frame.

The fundamental issue, then, is the lack of a common currency (denominated in terms of shareholder value added) that can be used by management to assess the company's entire investment opportunity set. Traditional financial techniques such as DCF and decision trees are applicable only to a small subset of such opportunities. The reason is that the constraining assumptions in DCF and decision trees—namely, that cash flows are "deterministic" and that there is thus no decision flexibility in the future—do not hold for most investment opportunities in R&D. Although decision makers may ask that such traditional analyses be conducted on a larger number of opportunities, they generally know that the results will not be sufficiently robust to make decisions. In such situations, smart decision makers will be more interested in the assumptions used by the analyst and less in the results of the analysis.

To remedy this situation, we need methodologies and tools that satisfy the following criteria:

1. The methodology is sufficiently generalized to be applicable across the entire investment opportunity set.
2. A tool is available that can be consistently and systematically used across all opportunities.
3. Application of the methodology and the tool is as fast and easy as the application of traditional techniques such as decision trees.
4. Senior decision makers understand both the advantages of the method and the need for change.
5. Application of the tool is sufficiently systematic to be repeatable throughout the organization.

To analyze all investment opportunities in an enterprise, we need a flexible methodology that allows problems to be specified that have both private and market risks. In pharmaceutical R&D, private risks are related to experiments testing the safety and efficacy of the candidates. There may be private risks in R&D manufacturing as well, leading to nonscalability, lack of stability, or inability to manufacture within certain cost thresholds. Such risks should be treated separately and differently from market-related risks. Market risks have to do with the anticipated revenue streams from those products that end up passing the technical hurdles and are actually brought to market.

Another important consideration is that the follow-on capital investments required at different stages of the experiment and the duration of the experiments are highly uncertain at the outset. Nevertheless, there are large amounts of data from repeated and standardized experiments run in pharmaceutical R&D that enable companies to form probability-weighted expectations—and such expectations should be used in the analysis instead of the averages typically used in DCF and decision tree analyses. Because of the focus on averages in traditional analysis (average cost, average time, average revenue), significant time and effort are devoted to getting these averages "right." But, as experienced managers and analysts recognize, the averages that result from such analyses generally obscure more than they reveal. Given the variability and uncertainty associated with the costs and benefits of R&D programs, the best one can do when contemplating a new program is to start by providing a good description of the range of possible outcomes.

Decision Options methodology is ideally suited to tackle problems encountered in the life sciences business. Using decision tree-like constructs to represent managerial choices in response to variability in expected timelines, costs, and revenues enables the analyst to distinguish between flexible and committed decisions and between the effects of private risks and market risks. Since the method is not specific to any situation or set of assumptions, it is applicable across the entire investment opportunity set in the company. Moreover, by adopting such a framework that focuses on uncertainty rather than average outcomes, companies can actually reduce both the amount of data and the mechanical intensity of the valuation process.

Let us consider an R&D program in detail. Pharmaceutical R&D is done in stages. It is a complex process that requires many disparate functions to work together to advance an idea to market. The major stages are shown in the table, and approximate durations of these phases are also given. The journey from idea to launch takes over a decade, and it is a grueling process of trial and error.

Stage	Duration (years)
Idea	2
Lead	2
Candidate	2
Phase I	1
Phase II	2
Phase III	3
Registration	1
Launch	

What the company gets at the end of this decade-long process is a marketable drug for a specific disease or class of diseases. That product is a tradable asset (just like a stock), and this chapter's discussion assumes that all life sciences assets follow the geometric Brownian motion (GBM) stochastic process. Each of these phases can be further divided into subphases and may have many decision points. Further, many different organizations have to make decisions regarding manufacturing, testing, and marketing of the drug during the process, and their decision processes may or may not be fully aligned with the product timeline shown here. All these complications are addressed as we dig deeper into the life sciences R&D process.

The following are some examples of the type of investment decisions that a pharmaceutical company makes routinely.

1. Buying or selling an intellectual property (IP) position (R&D candidate, patent, proprietary technology etc.) from another company such as a biotechnology company
2. Investing in an internal R&D program, including money, people, space, manufacturing capacity, tax credits, and so on

3. Collaborating (and investing) in an external R&D program with another company, university, or the government

4. Selecting projects to form a portfolio (with or without a constraint on available resources of money, people, skills, space, capacity, etc.)

5. Designing a project plan for an R&D program (when, what, how, and how much)

6. Making insourcing and outsourcing decisions (where and when)

7. Making decisions around adopting new technologies and abandoning old technologies (when and how)

8. Making decisions on the nature of experiments (how much information to buy and when)

9. Making project selection/design decisions and marketing decisions when competitive information is present (label design, launch timing, and pricing)

10. Infrastructure decisions (buildings, space, locations)

11. Organizational design, compensation, and incentive decisions

The pharmaceutical business currently is largely based on creating chemical agents to effect a specific action in a complex biological entity. These new chemical entities (NCEs) can be designed to have specific actions on cell surface receptors that modulate the cellular signaling pathways controlling cellular function. *Pharmacology* is the science of how drugs affect biological entities. It has two aspects: pharmacokinetics (how the body affects the chemical) and pharmacodynamics (how the chemical affects the body). Since the NCE is "new" by definition, its pharmacokinetic properties (absorption, distribution, metabolism, and excretion) are not precisely known. The manufacturing process of the NCE also has to be invented from available chemicals (some of these processes can have as many as 15 manufacturing steps of reactions starting from what is called alpha raw materials (common chemicals). Each manufacturing step represents the manufacturing of an intermediate chemical from existing ones, which then feeds into the subsequent step. Also, one can only guess the pharmacodynamic properties of the NCE and its hypothesized effect on cell receptors and how such effects ultimately translate into efficacy (beneficial effect) and toxicity (bad effects). Often, the difference in efficacy and toxicity (therapeutic index) determines the viability of the drug as there is no chemical without some toxic effects (side effects) and none with perfect efficacy.

Discovery and development of a drug include the manufacturing of a new chemical, the properties and manufacturability of which are not precisely known. The chemical has to be tested in complex biological systems, but the effects of the chemical cannot be precisely predicted. This means that all through the R&D process we are dealing with various types of uncertainties, and the decisions regarding the nature, timing, and size of manufacturing and testing activities have to be made in a highly uncertain environment. These uncertainties can be largely divided into the following four categories:

1. *Cost uncertainties:* The cost of conducting an R&D program depends largely on the people and materials needed to manufacture and test the prototype. The prototype here typically is an NCE or an NDE (new device entity). The "people costs" (largely driven by compensation) may represent over 75% of the costs in this people-intensive business, and the rest is related to materials, equipment, and other infrastructure. A typical R&D program involves various specializations such as clinicians, toxicologists, engineers, chemists, biologists, statisticians, and others. The complexity of the program as well as the company that conducts the program also requires support personnel, including human resources, information technology, infrastructure maintenance, accounting, and management. So, in addition to the "direct" costs of people involved directly in the production and testing of the prototype, we also have to include "indirect" costs related to the support personnel as well as infrastructure.

   Costs thus have a variable component as well as a fixed component. Fixed costs represent infrastructure costs and some of the support costs that cannot be avoided in the short run. R&D programs are also conducted in stages, so costs have to be estimated for each stage separately. In each stage, certain quantities of the NCE are manufactured (called drug substance). The drug substance is then formulated into a drug product (such as a tablet or capsule) that can be used in an experiment. In certain animal experiments or early human experiments, the drug substance could be used directly without formulation. In the manufacturing of the drug substance and subsequent formulation into drug product, one has to estimate the quantity needed for the current experiment and possible future experiments. Further, since the manufacturing and formulation processes have to be "invented," the precise yield (the amount of finished goods as a function of raw materials) is also not known. In making decisions regarding a specific R&D program, one should only include costs that are "avoidable" (or marginal). This is a function of the nature and timeline of the program, and since one cannot precisely predict the timeline or the complexity of manufacturing and testing *ex ante* (because of all the uncertainties discussed), one can only get an uncertain estimate of the costs of the proposed R&D program.

2. *Timeline uncertainties:* The R&D programs in pharmaceuticals are discovery-and-development programs. They start with a hypothesis regarding how a disease can be treated, cured, or prevented by effecting certain actions in the body by introducing a chemical entity

in a specific fashion. Since none of the components of this hypothesis are precisely known, the manufacturing of the drug substance and drug product for conducting the experiments requires the creation of a new manufacturing process. This process requires starting with raw materials that may be purchased from many vendors around the world or only a few specialized ones. Uncertain lead time exists between the order and arrival of starting materials from a vendor (who has to deal with other sets of uncertainties related to orders from many customers, availability of equipment and raw materials, shipping, and regulatory clearance). Once the starting materials arrive, further manufacturing delays may follow related to availability of personnel and equipment, availability of the specification of the manufacturing and quality control/testing processes, and the design and configuration of the manufacturing kit. Since the yield from each manufacturing step is not known, steps may need to be repeated to manufacture the necessary quantities. Since the impurities in the manufactured drug substance have to be precisely controlled (as impurities may have unanticipated toxic effects not related to the NCE itself), certain batches of the manufactured final drug substance or intermediates may have to be discarded if found to contain unknown impurities or known ones outside control limits. All of these possibilities make the timeline for the manufacturing of the drug substance and drug product to the required quantity highly uncertain. To make matters more complicated, there are compounding timeline uncertainties on the testing side. Experiments have to be run on animal and human models in stages. In animal models, availability of specific kinds of animals as well as personnel to conduct the experiments may introduce delays. In human experiments, selection and establishment of investigator sites (clinicians who conduct the tests), enrollment of subjects or patients (people who volunteer to undergo the study), and unanticipated and unique side effects (sometimes resulting in the shutting down of the experiment in a person, site, or the program) all can introduce uncertainties in the timeline.

3. *Success rate uncertainties:* In each stage of the R&D program (and virtually at any time), new information arrives from experiments that may result in a reestimation of the anticipated success rate of the program. For obvious reasons, the R&D program will proceed only if the expected success rate is high enough and if the expected profits from manufacturing and marketing of the drug are high enough. We use "high enough" here as traditional decision processes are based on "qualitative metrics" and generally not on a holistic shareholder value metric. Two types of "bad news" could force a "rethinking," "slow down," or even "abandonment" of the R&D program.

The first one is the lowering of the therapeutic index, either due to lower-than-expected efficacy (the drug does not produce the beneficial effects anticipated) or a higher-than-expected toxicity (the drug has higher-than-expected toxic effects) or both. Such information arrives from the animal tests or clinical experiments conducted. The second type of bad news is that the cost of manufacturing the chemical is substantially more than expected due to higher-than-expected raw material costs, lower-than-expected yield, or higher-than-expected equipment costs. There can also be "good news" (the opposite of events described for bad news), which can result in a decision to "accelerate" the program by increasing the enrollment rate of patients and subjects by increased incentives. In either case, arrival of new information may force us to change our expectation of "success" of the R&D program. Most often, the success rate is considered to be a binary outcome; it either succeeds or not. But, it is not always the case that the program suddenly dies. New information may force a redesign or modified label expectations (what can be sold in the market).

4. *Market potential uncertainties:* Pharmaceutical R&D is a long-drawn process, often taking over a decade from idea to market. For many decision makers in R&D, this is a long time, and considering the market potential in R&D, decisions are often difficult. However, market potential is obviously an equally important component of R&D decisions. The expectations around market potential are unclear when the program starts and continue to change all through its progression, affected by competitive and regulatory actions, label expectations, therapeutic index, and a host of other factors.

In making a decision in R&D, all one can do is to capture all the uncertainties and use them systematically. Existence of a complex set of uncertainties should not be the cause of adoption of *ad hoc* decision processes. Shareholders do not pay decision makers to gamble; they expect them to improve decision quality. Decision quality is related to the systematic use of all available information.

Project Design under Multiple Uncertainties

Let us look at a stylized problem in R&D. Figure 9.1 is a three-stage R&D program. The three stages are represented by costs, timelines, success rate, and a gain at the end of the decision options tree. The first two stages require an experiment that results in a technical outcome. In this case, a

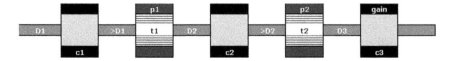

FIGURE 9.1
Representation of a three-stage R&D program.

technical outcome is a binary event that makes the drug technically viable or not. If the drug is not technically viable (either because its therapeutic index is low, i.e., low efficacy and high toxicity, or manufacturability is not possible for a reasonable cost), it has to be abandoned. If the drug "passes" and subsequent investments are made, it will eventually create a gain at the completion of stage III. At the end of stage III, it can be sold for the anticipated "gain" to another company. If the company does decide to go forward at any stage (in essence, exercising the option to invest in a stage), each of the stages also has uncertain costs, represented by $c1$, $c2$, and $c3$. After investing in each stage (investments are taken in manufacturing and testing of the drug), we get information that results in a binary outcome: the drug is technically viable or it is not. We represent these by probabilities $p1$ and $p2$. In stage I, there is a probability $p1$ that the drug will "technically succeed." In other words, there is a probability $(1 - p1)$ that the drug will fail for technical reasons after stage I. Similarly, there is a probability $p2$ that the drug will technically succeed after stage II. Each of the stages also has uncertain timelines; we cannot precisely predict how long it will take to manufacture and test the drug (for all the reasons mentioned).

As you can imagine, this indeed is a complex problem—decisions taken in a sea of uncertainty. Decision makers in the life sciences industry struggle with similar problems every day. Not a single day passes without new information that may force them to rethink their strategy. We simplified the current problem by constraining it to a few stages and few decisions. In actuality, there are hundreds of these types of decisions in the progression of a drug from idea to market—a journey that takes more than a decade. Decision complexity does not abate after the drug is approved and reaches the consumers for the first time. In some sense, it is only a start. Every time a patient takes the drug, more information is generated regarding the drug's safety and efficacy. Unanticipated adverse events do happen (some of these may be due to the specific attributes of the person taking the drug and some due to interactions with other drugs taken or a complex combination), and this may prompt the Food and Drug Administration (FDA) to reevaluate the drug. This may involve stricter warning labels, requiring the drug to be given only in the presence of a doctor or in a hospital. It can also result in a withdrawal of the drug from the market altogether. So, the history of any drug is fraught with uncertainty and the arrival of new and unanticipated information over multiple decades.

dostoch	init	mini	maxi
gain	700	500	1000

doother	type	aver	stdv
c1	cons	15	
c2	logn	300	50
c3	logn	300	75

doother	type	aver	stdv
t1	logn	1	0.25
t2	logn	2	0.5
p1	cons	70%	
p2	cons	80%	

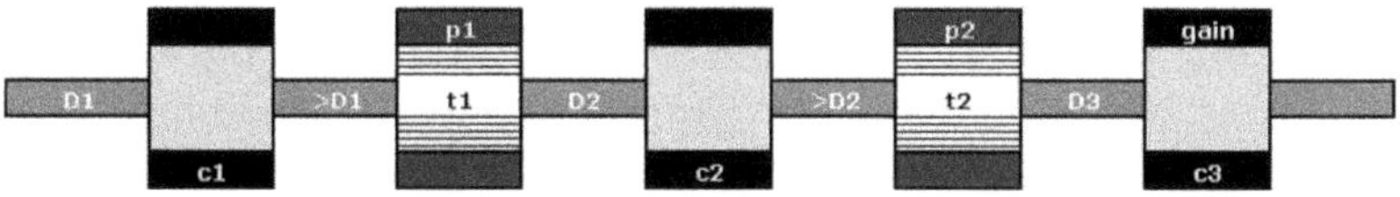

FIGURE 9.2
Modeled uncertainties in a three-stage R&D program.

Let us model some of these uncertainties (Figure 9.2). First, assume that we have done thorough market research on the gain we can expect if the drug completes the three stages of development. This analysis may have taken into account the overall market for the drug, including patients seeking treatment; prescription patterns; compliance rates; existing drugs of similar types; reimbursement patterns (since the insurance companies typically pay for the drug), including the probability of being in the formulary (list of drugs approved by insurance company); and many other factors. None of these is precisely known, but we can get imprecise estimates of all these factors. Using these estimates, we can calculate what the drug's value could be at the completion of the three phases. This is the price we can expect to get in an external transaction with another party and is in fact the NPV of the free cash flows generated from the drug (assuming it is marketable at the end of the third phase). The free cash flow from the drug will be a function of total units that can be sold, price, the growth of units over time, the remaining patent life (discussed separately in this chapter), the cost of manufacturing and logistics, marketing costs, and other general and administrative costs.

The remaining patent life of the drug may need explanation. Since it is an NCE, the company may have taken patents on the composition of the chemical, the method of manufacture, the mechanism of action (how the drug acts on the body), and possibly other attributes of the therapy. These patents, once issued, are given for a fixed period (approximately 20 years) from the date of issue. Once the patent life runs out, the drug becomes *generic*, allowing any company to produce and sell it. If this happens, the premium pricing enjoyed by the inventing company disappears, and the drug's market price will drop quickly. Generic pharmaceutical companies that dominate the segment of such generic drugs have no R&D costs, and their cost structure is similar to commodity chemical companies. Once the manufacturing process of the drug is known and the inventor of the drug has no exclusive right to it, generic companies simply can replicate the process, produce the drug, and sell it under a different generic name (the original company may still have the trademark on the original name). The generic market is increasingly competitive, so the loss in sales of a drug after a patent runs out ("loss of exclusivity") is indeed dramatic.

The "gain"—the value of the drug as it enters the market—has to consider all possible cash flows from the product before and after loss of exclusivity (end of patent life). The "expected free cash flows" need to be discounted back in time using a discount rate commensurate with the systematic risk of the drug. Note that many companies use only the weighted average cost of capital (WACC) of the company, and this is incorrect. Since the company engages in a variety of projects, the use of WACC in the discounting of a single project can either underestimate the project's risk (resulting in a higher-than-appropriate NPV) or overestimate the risk (resulting in a lower-than-appropriate NPV). Also, note that a straight application of discounting based on the capital asset pricing model (CAPM) is sufficient here as once the drug is ready to be marketed, there is no more flexibility associated with R&D decisions (the only type of decisions we are considering in this particular problem); thus, the free cash flow stream from the drug has no optionality.

By considering the various uncertainties described, we can get an average NPV as well as minimum and maximum NPV. We calculate the minimum NPV by considering a scenario in which the cash flows are the lowest (low revenue, high costs, and short period of time to loss of exclusivity) and the maximum NPV by considering a scenario in which the cash flows are the highest (high revenue, low costs, and long period of time to loss of exclusivity). The NPV is the "gain" in the decision options problem. We assume that this is a traded asset; thus, its value follows GBM. We further assume that this gain can be replicated in the market, so the value of the decision options (which are options on this underlying asset) can be calculated using the risk-neutral valuation method. Options purists may object to it by arguing that the gain cannot be perfectly replicated in the market. Every time we come back to this question, think about the following:

1. To employ risk-neutral valuation, we do not need a "perfect clone" of the underlying asset. The asset only needs to be "replicatable." The gain may be replicated in this case by using the stock of a biotechnology company with a project similar to the one under consideration. Or, it could be replicated by a complex combination of securities such as exchange-traded funds that include many companies that focus on a specific indication (as targeted by the drug). Such a bundle may also include short or long positions in futures such as the futures traded on presidential candidates if the pricing is affected by certain political parties remaining in power or not.

2. This replication process is "imperfect." The market portfolio used in a single-factor model such as the CAPM is also not perfect; the calculation of β using a market-traded proxy is also imperfect. The idea here is that the framework is a reasonable representation of what happens.

3. In a world with a large number of participants and increasing level of information flow, assets will continue to move toward a state at which replication is always possible by a combination of traded assets by a wide variety of participants, including investors, employees, suppliers, and customers of the company.

This means that we now have an asset called *gain* that can be assumed to follow GBM, and we can employ risk-neutral valuation. We also know the current value of gain (which is the average NPV) as well as the minimum and maximum value. Using these three (average, minimum, and maximum), we can create a "volatility" parameter of gain that is assumed to follow the GBM price process. As we have seen, volatility is a measure of how much the price of an asset (in this case gain) moves up and down during its lifetime. Since we know how GBM progresses (thinking back to the stochastic process equations) and the minimum and maximum values it can take within a period of time, we can back-calculate the volatility. Note that we are not doing any inflation adjustments, so all numbers are real (not nominal).

Shown in Figure 9.3 is the risk-neutral simulation of "gain" over approximately 5 years. As we will see in future problems, there may be other types

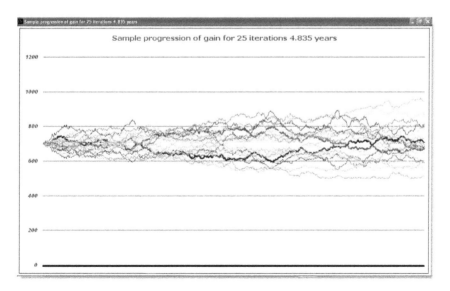

FIGURE 9.3
Stochastic simulation of gain from the R&D program.

of options present even after the drug is ready to be marketed, such as the following:

1. Option to expand to other geographic areas. Typically, the approval of the drug is sought in one geographic area (such as the United States or European Union) first and then expanded to other regions.

2. Option to expand the "label" (the description on the drug describing the type of diseases [indications] for which the drug is effective). Typically, the drug is first proven to be effective for one or a few indications before it is brought to market. The company may try the drug for other indications at a later time.

3. Option to create different formulations and dosage forms. Typically, a single formulation and dosage form are used at the market entry. The dosage form is the drug product, such as a tablet or capsule. The company may choose to improve the drug's characteristics with better dosage forms in the future.

We ignore these and other postmarketing options for now.

Next, let us look at costs. We have three costs ($c1$, $c2$, and $c3$) that represent the three stages of the R&D program. We are embarking on stage I immediately, so we have a really good estimate of what it will take. In fact, we are so sure of this, we are going to assume it is a constant: $15 million. However, $c2$ and $c3$ are going to occur at some future point in time, so we only have some expectations for them. We will also only know how much these costs are when we start those stages. For example, we assume $c2$ will be around $300 million. By analyzing historical data of similar projects, we anticipate that $c2$ will be a sample from a lognormal function with an average of $300 million and a standard deviation of $50 million. Given in Figure 9.4 is the distribution for $c2$. This means that the cost for stage II is in the range of $180 to $480 million, with an average of $300 million.

As is typically the case for cost estimations, there are small probabilities that we will have significant cost overruns (the distribution has a positive skew and fatter tails than a normal distribution). There are other types of distributions (such as γ) that may be used to better describe cost overruns, but throughout this book the concept is kept simple, using only one of the three common distributions: normal, lognormal, and triangular. Lognormal distribution is always used to represent time and cost overruns in project plans. Similarly, we assume the expected cost of stage III is also $300 million, but it is little bit more uncertain, showing a standard deviation of $75 million. The distribution of stage III cost is given in Figure 9.5. The range of expected stage III cost is $140 to $570 million.

Now let us study the timelines for the project. Here, we are assuming that we commit some budget when we start each of the stages (represented by $c1$, $c2$, and $c3$) and the project timelines for stages I and II are $t1$ and $t2$,

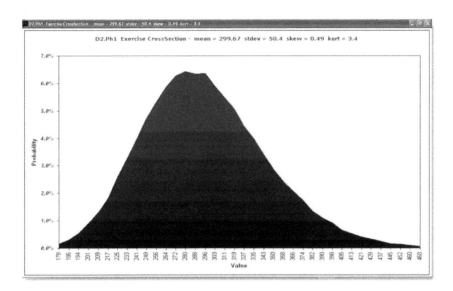

FIGURE 9.4
Probability distribution of cost in stage II.

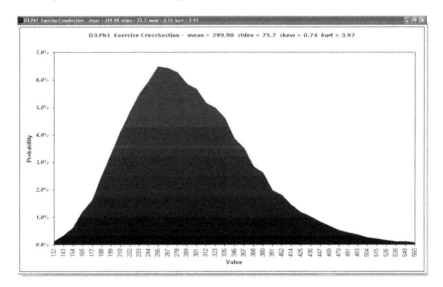

FIGURE 9.5
Probability distribution of cost in stage III.

respectively. Stage III happens immediately after stage II, and we have the option to buy gain by spending $c3$. We can imagine this as committing the cost of conducting stage III in a contract with the buyer of the drug at the end of stage II. The selling company (the company conducting stage I and stage

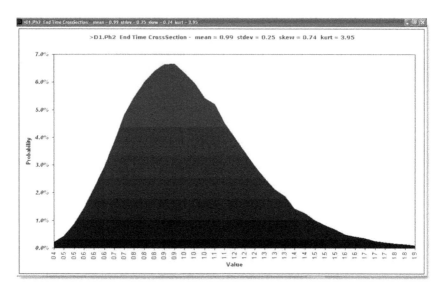

FIGURE 9.6
Probability distribution of time in stage II.

II experiments) is also committing the expense of conducting stage III but selling the drug at that point to another entity.

Note that we have to commit the stage I cost of $15 million immediately. This is a cash flow, not an option as there is no time to make this decision. We assume the time that will be taken to conduct stage I is approximately 1 year. However, we do not know this precisely. Analyzing past experiments of similar nature, we estimate that the timeline is a lognormal distribution with expected outcome of 1 year and a standard deviation of a quarter (0.25 year) as shown next. This means that we can get really lucky and complete stage II in 6 months (0.5 year), or it could drag on for as much as 2 years. Figure 9.6 shows the probability distribution of the stage II timeline.

The only remaining item that needs discussion is the success rate. Success rates are sometimes called private risks (to differentiate them from market risks), and they are not correlated with the market. For the current case, we are going to assume that the risks are binary outcomes. For example, we are 70% confident that a stage I experiment will be successful. This may be based on historical data or expert judgment. For example, we may find that 30% of similar experiments in the past failed, resulting in a complete stop of the R&D program. We call these types of risks *binary* as they are either good or bad (and nothing in between). In many cases, the outcomes may not be as clear-cut, and they may result in either diminished or enhanced expectations around the technical aspects of the product (and its marketability). We consider such risks in other cases.

Now that we have defined the asset (gain from selling the drug at stage III); the costs of conducting the program, including a committed cost of the stage III program (to be conducted by the buyer of the drug); the timelines for the different stages; and the success rates, we can calculate an economic value of this decision options bundle.

To value the current decision options problem, we need one more item: the risk-free interest rate. Remember that we are going to value this in the risk-neutral framework (taking advantage of the assumption of complete markets). Also note that while calculating expected costs and gain (NPV of cash flows), we used non-inflation-adjusted real numbers. This means that the risk-free rate has to be the real risk-free rate. A 5-year T-bond (which is the approximate duration of this decision options problem) yield gives us the nominal risk-free rate. By subtracting the inflation rate from it, we can calculate the real rate. For convenience, we assume that the risk-free real interest rate is 0. For most of the problems of this nature, we continue to assume a real risk-free rate of 0 as it is close to what we currently observe. It is possible to use a term structure in the risk-free rate as the yield curve will be sloped, but for most practical problems, such complexity does not improve decision making and may take the decision maker into long-drawn discussions around theoretical nuances that have little effect on ultimate conclusions.

Analyzing the current problem, we find that the value of this R&D program currently is $13 million. So, the company should go ahead and invest in stage I. Figure 9.7 shows the completed model.

Now consider that we have an opportunity to conduct stage II differently. We can break it into two; call these stages IIA and IIB as shown in Figure 9.8. Each of these substages consumes 50% of the cost of stage II. So, in combination,

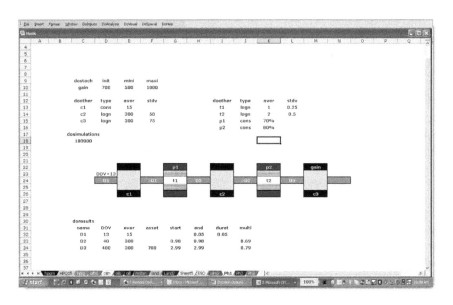

FIGURE 9.7
Completed model of a three-stage R&D program.

FIGURE 9.8
R&D program with split stage II investment sequence.

they still total the original cost of $c2$. It is just that we are breaking the experiment into two smaller ones. It is also the case that the time taken for conducting each of these substages is 50% of the original stage II timeline. So, again in combination it takes the same amount of time to complete the entire stage II. Also assume that the success rate from stage II remains the same. Remember that we had an original stage II success rate of 80%. This means that 20% of the time a drug failed in stage II. To have the same success rate from stage II, the success rates from each of the substages have to be higher than 80%. Why? If we assume that both stage IIA and stage IIB have the same success rate as before (80%), then the combined success rate of stage II will be $(80\%)^2 = 64\%$. Remember that we are combining two probabilities in sequence here. So, the success rate in each of the two substages has to be $\sqrt{80\%} = 89\%$. Everything else remains the same. The only difference in this design is that we have broken up stage II into two parts with a decision point between. At that point (halfway into stage II), we can also make a decision (an option) whether we want to proceed further. By then, we may also have a better estimate of gain, having observed it for the past couple of years.

Intuition tells us that such a design introduces a higher level of flexibility into the decision process. With everything else (timeline, cost, risk, and gain) remaining the same, it is also likely to increase value. As shown in Figure 9.9, this additional flexibility increases the value of the R&D program to $16.6 million from the original $13 million, a 28% increase.

This is an important insight for those designing R&D programs. As we intuitively know, the more flexibility we have in the program, the higher the value of the program will be. It should also be clear to the user that the incremental value generated by the ability to walk away from the program at stage IIA. Such an "abandonment option" is not related to the failure of the drug in an experiment (technical failure) but rather to a reevaluation of the program's potential in stage IIA. Note that by the time stage IIA is completed we have completed $t1 + t2/2$ of the overall timeline for the program (approximately 2 years). During this time, we may have gotten more information on gain, including but not limited to pricing power, competitive R&D, government and regulatory policy changes, public attitude toward seeking treatment for the specific disease targeted by the drug, and more. This better estimate of gain may have resulted in a loss of NPV, and we may reconsider the original plan to invest in stage IIB and stage III. If gain has fallen to a sufficiently low value, it may force abandonment. It is this decision flexibility that enhances the R&D program value by 28%.

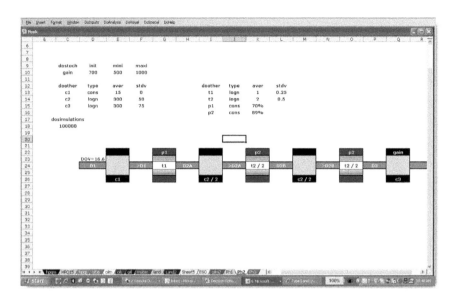

FIGURE 9.9
Completed model for split stage II R&D program.

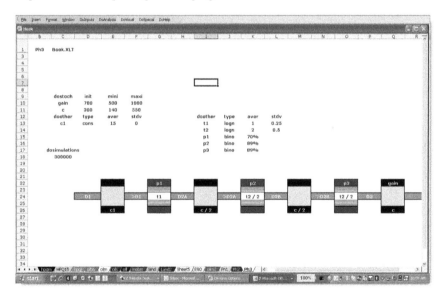

FIGURE 9.10
R&D program with correlated costs.

Let us make this problem a little more complicated. Assume that we can actually observe the costs of stage II and stage III over time, and they are correlated. We can model this problem by assuming cost also follows the GBM. The model is shown in Figure 9.10.

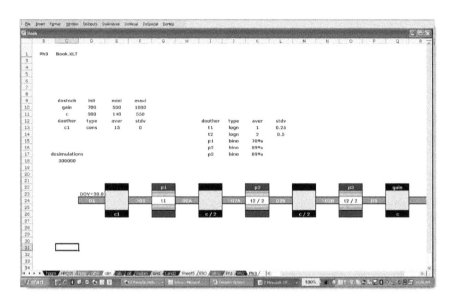

FIGURE 9.11
Completed model for R&D program with correlated costs.

In this case, we create a stochastic variable called c and use that in stages IIA, IIB, and III. Since we are using the same variable, we are also assuming that the costs in these stages are perfectly correlated. The lower stage II costs are, the lower stage III costs will be and vice versa. As we go forward in time, we get better and better estimates of the overall cost of the program (Figure 9.11). In this case, at the completion of stage IIA, we not only get a better estimate of gain but also better estimates of all costs for all subsequent phases as they are correlated with the costs just incurred in stage IIA. Intuition tells us that this may further increase decision flexibility as the decision to abandon in all stages after stage I can now be based on better cost estimates. Since the company is interested in the highest profits, the cost information combined with the gain information is more powerful in making abandonment decisions.

We find the value of the R&D program to be approximately $31 million, nearly double the case when the costs can only be known when a stage started and costs from previous stages do not include any information regarding costs in subsequent stages. In calculating the R&D program value, such assumptions can become critical and will have meaningful effects on decisions, especially when all feasible R&D programs cannot be pursued due to capital constraints, and the company is forced to pick a limited set of programs to pursue.

You may also have realized that experiments that reveal information such as cost and gain have value. For example, one could visualize a pilot manufacturing process experiment that can be conducted quickly (and cheaply) to get a better assessment of costs prior to starting the large experiment in

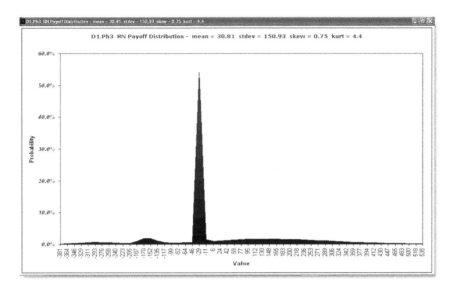

FIGURE 9.12
Risk-neutral payoff distribution from the R&D program.

phase II. Similarly, the company could interview doctors to better assess the drug's potential if it were to be successful in R&D.

One assumption with which options purists may take issue is the idea that costs follow the GBM with a risk-neutral drift. Here, we treated costs exactly like gain, so all assumptions regarding replicability now have to hold for cost also. How can one replicate costs? Perhaps this can be done through traded assets such as the stock price of contract manufacturing organizations (CMOs) or contract research organizations (CROs) that may have entered into fixed-price contracts with pharmaceutical companies. Since costs are a function of supply and demand of materials, manufacturing capacity, and patients, CMOs and CROs may also be driven by the same dynamic. In any case, remember that we are not really trying to precisely replicate the underlying R&D program but doing a mental exercise on how replicability can be achieved. We can also visualize the risk-neutral payoff distribution from this R&D program (Figure 9.12).

The peak in Figure 9.12 represents the high probability of abandonment at stage IIA. Note that this distribution is not the real payoff from the program (as we are using a risk-neutral valuation technique to value it). But, the shape of the payoff is instructive and demonstrates reasonably high levels of possible losses (albeit low probability) from pursuing the program. Some of these are due to technical failures in the various stages. After investing in the program, we find out that it does not work (a technical failure). Some losses are due to abandonment since after investing in some stages, we have decided to cut the losses as the cost rises or gain diminishes. If the program succeeds, there are also high gains, as shown in the right tail of the distribution. These big gains make up for all the losses, especially because of a few relatively high-payoff scenarios in which we get lucky with the gain (highly valuable drug) or our costs are a lot lower than anticipated and the program progressed rapidly. Note that we are ignoring the changes in the period of loss of exclusivity due to a faster or slower program. In future cases, we demonstrate how this effect can also be incorporated.

Decision to Invest More in an Existing R&D Program to Enhance Success Rate

One of the questions with which R&D managers often struggle is whether it is worthwhile to invest more into an ongoing R&D program. Consider a decision to buy information that increases the success rate of a program. Information is "bought" by investing more into the program either by running a larger clinical study or by creating a better drug product that improves the action. It can also be other type of investments such as equipment, internal or external personnel, and adding other types of studies.

Improvement in the success rate has to be understood in the context of errors in technical decisions. The technical go/no-go decisions are typically made on a therapeutic index such as the therapeutic ratio (TR), which is the ratio of the dose of the drug that creates a toxic effect for 50% of the study population (T_{50}) and the minimum effective dose that is needed to create an efficacious effect in the same population (E_{50}). Obviously, the higher the toxic dose, the higher the TR, and the lower the efficacious dose, the higher the TR.

$$TR = T_{50} / E_{50}$$

The difference between a toxic and an efficacious dose can be very narrow (the therapeutic range or therapeutic window), so assessment of these doses is critical for making a decision to go forward with the drug after a stage is completed. In each stage, the drug may be tested in a certain number of subjects or patients. For example, in phase I, healthy volunteers (subjects) take part in the experiment for the assessment of the toxicity of the drug. Similarly, in phase IIA a group of patients goes through the experiment for the assessment of the efficacy of the drug. Since all human beings are not the same, the effect seen in one will differ from the effects seen in others and thus the information collected is not precise but probabilistic. Generally, the more information available (larger number of subjects and patients), the higher the confidence in the assessment of the TR and the higher the chance of making the "correct" technical go/no-go decision.

Clinical study design has to take into account the null hypothesis (which may be accepted or rejected) and the sample size needed to reach a statistical conclusion. Students of statistics are familiar with type I (false-positive) and type II (false-negative) errors. Since we start most experiments with a hypothesis and collect sufficient information to either accept or reject the null hypothesis, both of these types of errors are costly mistakes for the company conducting R&D. If type II error occurs (incorrectly concluding that the drug is not beneficial when it actually is), the R&D program may be abandoned. This is the error of accepting the null hypothesis when it should have been rejected. In this case, a potentially good drug will be abandoned, and the investments made up to that stage become a waste; more important, the company will "throw away" a valuable prototype. If type I error occurs (incorrectly concluding that the drug is beneficial when it is not), the R&D program will progress to subsequent stages, forcing the company to perform larger and larger experiments. This is the error of rejecting the null hypothesis when it should have been accepted and accepting the alternative hypothesis. As the "bad drug" progresses through the R&D process, "burning more money," more information will become available. This may result in the abandonment of the drug at a later stage. In some cases, the company may continue to make type I errors in all stages of R&D, taking the drug to market. However, once the drug is on the market, the number of patients taking it dramatically increases, and the information exponentially expands, making the error very transparent. Adverse effects may show up in some percentage of the population, and the actual therapeutic range reveals itself. Depending on the level of toxicity, a number of different outcomes—such as the FDA requirement of sterner warning, a black box label, hospital-based administration, or withdrawal from the market—are possible. Both type I and type II errors are extremely costly for the company.

Consider a comparative clinical study in which the null hypothesis is that the drug under investigation is the same as other existing therapies for the same disease. A type I error is concluding that the new drug is better than existing ones (when they are generally the same). In this case, the null hypothesis is falsely rejected. In any experiment, there is a probability that a type I error could exist (sometimes called the p value). Type I error can be reduced by increasing the number of data points (by having more patients in the study). However, it may also increase the type II error (the null hypothesis may be accepted that the new drug is not any better when it actually is). The type II error can be reduced by increasing the power of the study.

Note that type I and type II errors are also a function of the design of the clinical study. They are "statistical errors" assuming that the design of the study is correct in obtaining the information sought. For all the cases described in this book, it is implicitly assumed that the experiments are always designed appropriately, and that the type I and type II errors are low enough to be discarded. So, the success rate in a clinical study is the "true success rate," and we always make the right call around the technical success of the candidate given the information from the experiments.

Consider the R&D program given in Figure 9.13. The first section of this sequence of decisions deals with R&D as shown in Figure 9.14. This drug candidate is in phase IIB. Once that stage is completed, the drug will move into phase III. If that experiment succeeds, the company will have enough information to file for FDA approval (we call this stage REG for regulatory). There are expectations of cost for the three phases, designated by PIIBC, PIIIC, and REGC, respectively. There is also a placeholder for PIII for some "additional cost," designated as ADRC. Note that at Phase III, the final phase of this R&D program prior to submitting and NDA (new drug application), there are a number of trials already planned. As many as 2,000 patients may go through trials during this time (which extends 3 or more years). With the current design (which takes a total cost of PIIIC), we expect a certain success rate of PIIIp. In this success rate, there is a certain probability of type I and type II errors. As mentioned, we always assume that the "decision" is correct, and that the type I and type II errors are low enough that they do not figure in our decisions.

Based on historical data, the chance of success in this phase is estimated to be PIIIp. Similarly there are expectations of success rates for PII (represented by PIIp) and for FDA approval (represented by REGp). All of these success rates are binary. This means that from a technical standpoint, we assume that the program either succeeds or fails (no other options). Although this may be a reasonable representation for most advanced R&D programs in pharmaceuticals, it is not always the case. The information obtained from R&D experiments is not typically black and white, and decision makers have to analyze large amounts of conflicting information. In some cases, the company may find that although it is unprofitable to take the candidate forward, it may still be possible to sell it to another company that may have a lower cost structure, better expertise in certain areas, draw synergies from its own existing programs, or utilize its sales force more optimally. So, the decision may not always be a complete abandonment but allow other modalities such as out-licensing (selling to another company). In out-licensing, the company may also retain certain options to "take part" in the success of the program at a later stage when more information is available. Obviously, such options have a price associated with them.

The additional cost ADRC is for an additional clinical trial that will utilize a different type of dosage form. This "tablet" uses new technology that improves certain pharmacodynamic properties of the chemical. With this additional experiment (and cost), we can extract a new set of data that will not be available in the traditional study. In effect, we are trying out a "new way" to create the necessary effect from the drug candidate. It is complementary to existing design. We assume that this additional investment enhances the chance of success beyond the original PIIIp. It is possible that the new experiment has some effect on the overall type I and type II errors, but we ignore them for now.

FIGURE 9.13
Staged R&D program with revenues.

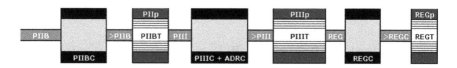

FIGURE 9.14
First section of the R&D model.

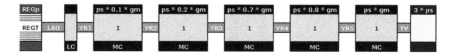

FIGURE 9.15
Portion of the model representing marketing.

The question we are trying to answer is whether the additional investment ADRC is worthwhile. To answer this, we can calculate the economic value of the R&D program with or without the additional investment. In analyzing the economic value, we also have to forecast what happens when the drug gets to market. The section of the model representing the market is shown in Figure 9.15.

Figure 9.15 shows the sequence of decisions after the drug is approved by the FDA. After the approval, the company commits launch costs (some costs may already be committed and invested while the drug is reviewed by the FDA as most companies do not want to lose much time after approval). The cost associated with launch is represented by LC. For the rest of the model, we assume that every year the company assesses the revenue potential for the drug and the marketing costs and makes a decision to keep it on the market or not. After 5 years, if the drug is still on the market, the company will sell the mature product to another company, and we expect three times the peak sales (represented by ps) in that transaction.

In every year, we calculate the revenue as a function of peak sales ps and gross margin gm. For the first 5 years, we assume a growth curve for the drug as a function of peak sales. For example, for the first year we assume 10% of peak sales, for the second year 20% of sales, and so on. We can create such growth curves from historical data of similar drugs. We also assume a marketing cost/year (represented by MC) irrespective of the sales. Since marketing costs are related to reaching doctors, insurance companies, and patients, they may be driven by other dynamics. We assume (in this problem) that they are uncorrelated with the revenue from the drug. We also assume that the gross margin, which is (sales − cost of goods)/sales, is a constant 75%. The cost of goods is generally well established by the time the drug is ready to be marketed.

Figure 9.16 shows all the assumptions related to status quo (the current design). In the current design, we have traditional clinical trials in both PII and PIII. Figure 9.16 shows the solution from the status quo analysis. Note that

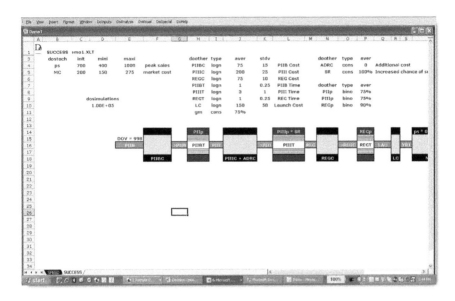

FIGURE 9.16
Completed R&D model.

we have set ADRC to zero and SR to 100%. This means that we are not taking any additional costs in PIII (for the specialized dosage form and modified study), and the success rate we expect is the same as calculated from historical data.

We have defined two stochastic functions: peak sales ps and marketing cost/year MC. Both of these can be observed over time, and we gain better understanding of both as the drug candidate progresses through the pipeline. We also assume that they are tradable assets, and that they follow a GBM price process. We also assume that both peak sales and marketing costs are "replicatable"; thus, we can solve this problem in the risk-neutral framework. The costs we assume to be probabilistic and independent. We assume all costs to be lognormally distributed, and we obtain this from historical cost data of similar programs. The duration of each phase, also obtained from historical data, is also represented by lognormal distribution. The success rates for PII, PIII, and REG phases are 75%, 75%, and 90%, respectively, and they are binary (success or failure).

We can visualize the risk-neutral price paths of both peak sales and marketing cost during the course of the decision options problem. Figure 9.17 depicts the price process for peak sales. Note that in many scenarios the peak sales hit the upper boundary of $1,000 mil (we assumed it is the maximum peak sales that can be achieved). This is a bit like introducing a mean reversion into the GBM price process. In this case, if the price hits the upper boundary, it does not go through it. There are many practical reasons why this is a reasonable assumption. A very successful drug becomes the target

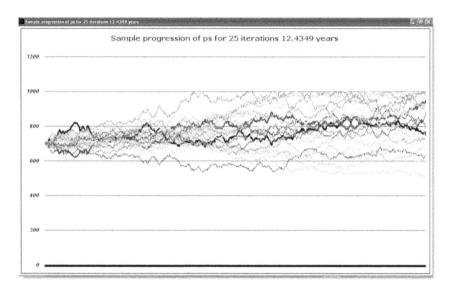

FIGURE 9.17
Stochastic simulation of peak sales.

for competition, which may bring similar products to market (acting as a curb on the upper limit on peak sales). Policy makers and insurance companies (which pay for the drugs) also take notice when a particular drug becomes very successful and may introduce limitations on pricing power.

Figure 9.18 shows the price process for marketing costs. This also has upper and lower limits. The upper limit may be self-imposed by the company, and the lower limit may happen when the company is obligated to conduct certain programs for education even if the drug "sells by itself." We also included the postmarketing studies the company may conduct (sometimes called phase IV costs) into the marketing costs. Any general and administrative costs (incremental to existing costs) are also included in the same variable. Note that we are only interested in incremental costs and not existing costs in analyzing the value of the R&D program. The company may not be able to avoid some of these costs even if the drug sells itself without any marketing by the company.

Options purists may object to boundaries imposed on the GBM in a risk-neutral framework. They may argue that such boundaries, if they exist for the reasons I described, are in the real world and cannot be artificially imposed on a risk-neutral price process. I would request that you do not lose too much sleep over such technical complications. In this problem, we can also argue that the appropriate risk-adjusted discount rate is the risk-free rate as these have low correlation with the market. So, both the price processes and the payoff diagrams are close to what we should observe in the real world.

Shown in Figure 9.19 is the risk-neutral payoff from investing in the drug candidate. The two peaks in the left are related to abandonment decisions at the beginning of PIII or after it (before submitting for FDA approval). Most of

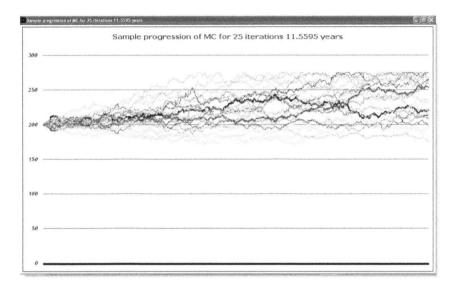

FIGURE 9.18
Stochastic simulation of marketing costs/year.

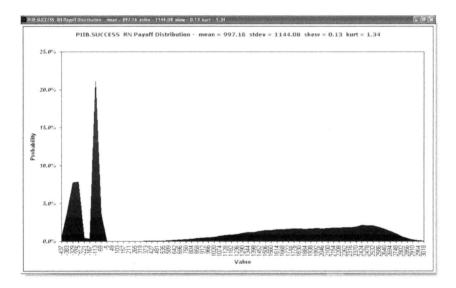

FIGURE 9.19
Risk-neutral payoff from the R&D program.

this is due to technical failures, and when the technical failure happens the company is forced to abandon the program and loses its investment. Note that the technical failure-related stoppage of the program is not an option. As often stated, an option is the right to do something but it is not an obligation. When

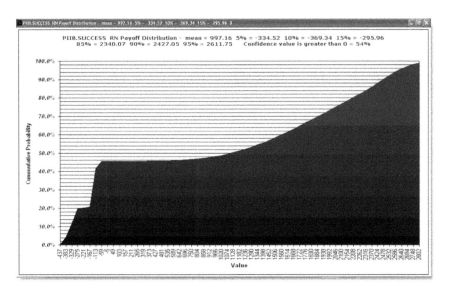

FIGURE 9.20
Cumulative probability of risk-neutral payoff from the R&D program.

technical failure happens, we are obligated to stop the program, but this is well compensated by the small probability of making large amounts of money if the drug succeeds and reaches high levels of peak sales.

The cumulative probability of the risk-neutral payoffs is shown in Figure 9.20. Although these are not the actual payoffs (as we are solving this problem in the risk-neutral world), they are instructive in understanding the risk inherent in engaging in this type of R&D. The first thing we notice is that there is only about a one-in-two chance of actually making money from this. Most of this is driven by private risk in R&D.

Now let us analyze the new plan in which we conduct additional experiments in phase III and conduct phase IIB in anticipation of it. This increases cost in phase III (but not in phase IIB). PIII costs increase by $50 million, and availability of results from this additional experiment increases the chance of technical success by 5%. We can now find the economic value of the modified program.

Figure 9.21 shows the assumptions and the results from the new design. We find that the $50 million additional investment is worth it. The economic value of the program increases to $1.05 billion. A 5% gain in economic value can be captured by the alternative design, which is more costly (by an additional $50 million in phase III) but improved the drug candidate's action. Note that we did not take into account any change in the duration of phase III due to the addition. We also did not explicitly model in the constraining patent life of the drug candidate, which will have an impact on the overall expected revenue if the time spent in R&D changes. We deal with these aspects in the next case.

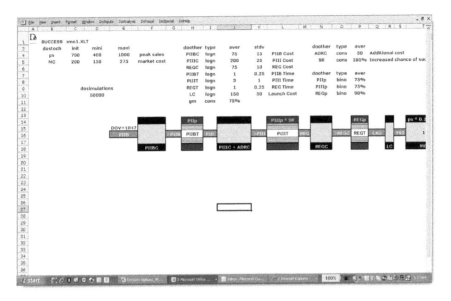

FIGURE 9.21
R&D program with early investments to reduce attrition.

Followers of pharmaceutical companies know that the risk of a drug does not go to zero after it is marketed. In many cases, we get more information that results in a loss of value for the drug. Competition is one reason for a spontaneous loss in value for a marketed drug.

Let us introduce a jump in the peak sales stochastic process to represent a spontaneous loss in value either due to competition or due to some regulatory or market change. Assume that there is a 2% chance every year from now that some adverse effect (competitive entry, regulatory change, etc.) happens, and if that happens, the value drops by 20% (but never below the minimum value of $400 million). Figure 9.22 shows the price process with these adverse events. The arrow in the figure shows a catastrophic value loss in one of the simulations.

Shown in Figure 9.23 are the assumptions and the economic value of the status quo program with spontaneous loss in value. We find that value declines to $950 million (5% loss). The 2% chance every year for such a loss does have significant effect on the economic value of the program.

The probability of technical success (a private risk) is an important attribute of value in R&D programs. In pharmaceuticals, nearly every program is new, and it is difficult to precisely assess the chance of success. It is always better to use a historical success rate in cases such as these as it is difficult for most to provide a probability of an event, especially if the event has binary (go/no-go) outcomes. The company's success rates may be more relevant than industrywide metrics. The historical success rate of the company at various stage gates is a function of expertise of the company, its business processes, and other natural factors that affect success.

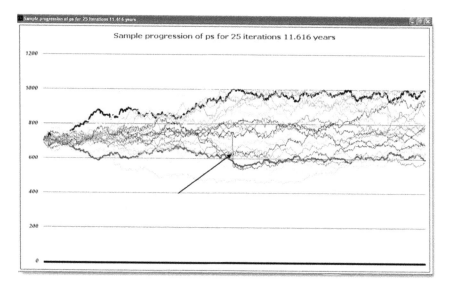

FIGURE 9.22
Stochastic simulation of peak sales with catastrophic value loss due to competitive entry.

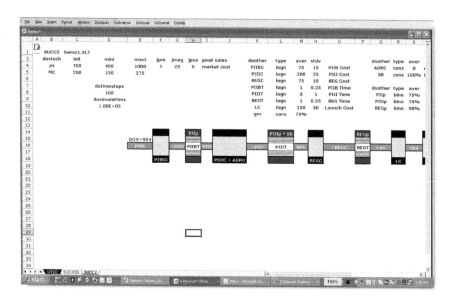

FIGURE 9.23
Economic value of the R&D program with competitive entry and catastrophic loss in expected peak sales.

Decisions to Accelerate an Existing R&D Program by Additional Investment

In the last case, we studied investing more assets into an existing R&D program to enhance the chance of success. In this one, we investigate how to evaluate the opportunity to accelerate an R&D program by more investing. There are many ways to accelerate an R&D program. Life sciences R&D requires many departments (e.g., manufacturing, clinical, and toxicology) to come together to conduct a program. One way to accelerate a program is to identify the department that is in the critical path and speed that up. For example, if manufacturing is in the critical path, it means that clinical and toxicology functions that use the drug to run experiments are waiting for the availability of materials. This delay can be avoided by finding additional suppliers or dedicating more manufacturing resources to the production of this particular drug. If, on the other hand, materials are sitting in the warehouse or at investigator centers waiting to be used and there are not enough patients to start clinical trials, incentives can be increased to accelerate enrollment, thus speeding up the R&D program. In either case, we can invest more into the department that is behind and tying up the process.

Consider the R&D program shown in Figure 9.24 for a drug candidate in PIIB. In PIII, we consider a scenario of investing more than status quo (that cost PIIIC) by additional cost (ADDC). The additional cost will be applied in the functional area that is in the critical path in the status quo project plan. If we do so, we expect to reduce the time needed for PIII by REDT (as in reduced time). On the marketing side, we have the model in Figure 9.25.

As in the preceding case, we have two stochastic functions, one representing peak sales ps and one representing marketing cost MC. We have a growth in peak sales reaching the maximum value in year 5. We assume that after the peak sale is reached, the company sells the drug to another firm, which may have a lower cost structure and can create a higher value from

FIGURE 9.24
R&D program in phase IIB.

FIGURE 9.25
Marketing model of the R&D product.

FIGURE 9.26
Terminal profits modeled as a stream of cash flows.

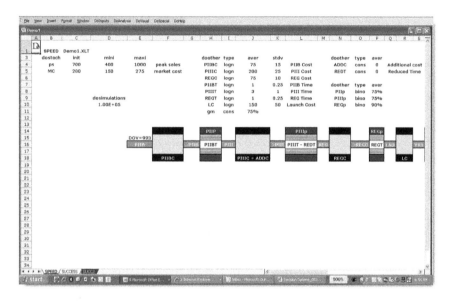

FIGURE 9.27
Analysis of the R&D program.

the mature drug. We also model the idea of a fixed patent life in the terminal value as shown in Figure 9.26. We assume that the drug currently has a patent life of 18 years. Since we have only uncertain estimates of the duration of the various phases, we subtract these variables from the total of 18 years (current time to exclusivity) to calculate the remaining patent life when the drug is considered mature and ready to be sold (after 5 years on the market). We also include the variable REDT as the time reduced from phase III as it will be an added benefit for the duration of the drug on the market. We assume that the terminal value equals peak sales × time to exclusivity × 0.4.

First, we calculate the value of the status quo project plan with REDT and ADDC set to zero. Given in Figure 9.27 are the assumptions and the result from the analysis. We find the value of the R&D program with status quo design is approximately $1.0 billion.

We can also look at the impact of our assumptions on the valuation. Figure 9.28 shows the impact of the peak sale assumptions on the economic value of the R&D program. Impact analysis shows that a 1% reduction in the average peak sale assumption will reduce R&D program value by 1.7%.

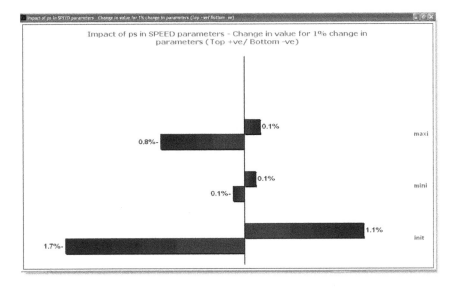

FIGURE 9.28
Impact analysis of peak sales assumptions on value.

Similarly, a 1% increase in average peak sales increases R&D value by 1.1%. The assumptions on the minimum value have little impact on the economic value. A 1% reduction in maximum value results in an 0.8% reduction in economic value of the program, but the impact is very little on the upside. These observations give us an indication of which assumptions are most sensitive to the results and may need more attention.

The new design with an additional $50 million invested to reduce the PIII time by 6 months (0.5 years) also can be valued to assess whether it is a better way to manage the program. Figure 9.29 shows the results from that analysis. The results show that the economic value of the program is roughly the same as before, with the new design marginally higher by 2%. Reduction of time in R&D has competing effects on economic value. The extension in patent life has a positive value as the drug can generate more revenue before it reaches loss of exclusivity. However, reduction in time also means that there is less time to observe what is happening to peak sales and marketing cost expectations before decisions are made to make further investments. Also, there is an obvious increase in cost ($50 million) due to the additional investments taken to accelerate the program. Note that the economic value increase calculated already has this additional cost considered, so it is better in this case to pursue the accelerated program. The accelerated program increases value by 2%.

Assume that this program has a 10% chance of encountering a negative event every year (such as competitive entry or regulatory change), and if that happens, value spontaneously drops by 25%. Figure 9.30 shows the analysis with the new assumptions incorporated in status quo. The economic value is

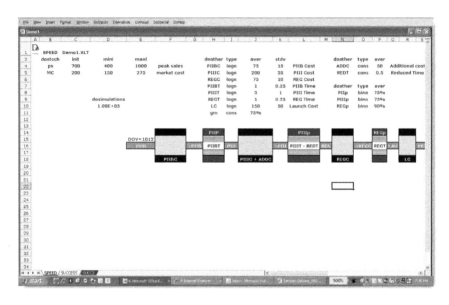

FIGURE 9.29
Analysis of R&D program with acceleration.

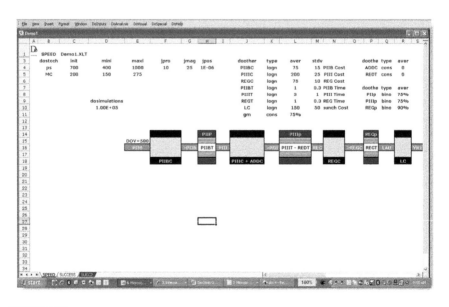

FIGURE 9.30
Analysis of R&D program with competitive entry.

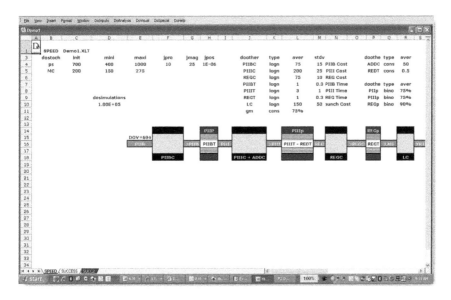

FIGURE 9.31
Analysis of the accelerated R&D program with competitive entry.

significantly lower, dropping to $580 million. A 10% chance every year is significant as the program duration is over 10 years. We can calculate the chance of the bad outcome not happening during the life of the drug as $(90\%)^{10} =$ 35%. This means that the 10% chance every year translates to a 65% chance of a bad outcome during the life span of the drug. The drop in economic value is over 40%. It is important to understand the probability of adverse effects such as these during the life span of the drug first and then translate that into a probability per year (as required in the model) to avoid mistakes.

Now let us analyze the accelerated program with the same assumptions. Figure 9.31 shows the results. The accelerated program is significantly better in this case. This is primarily due to the program reaching peak sales faster, reducing the overall chance of an adverse event during the high-revenue period.

Note that in all these cases we have not considered any pricing and marketing impact that may happen by reaching the market faster or slower. We also did not consider any competitive strategy. These effects can also be incorporated into the analysis. It is important that analyses such as these focus on describing the problem in rich detail with all associated uncertainties. However, any time spent to precisely define inputs (when uncertainty exists) is a wasted effort. The Decision Options framework allows the user to get comfortable with uncertainty and reach reasonable decisions faster using inputs with uncertainty embedded. The model described can be standardized and made available for decision screens within R&D. Such a process will allow the company to ensure that all operating decisions consider economic value systematically.

Nonbinary Risk in Product Enhancement

In previous cases, we modeled the private risks as binary risks. In many cases, this is a valid method. The information obtained from an experiment may result in a catastrophic failure of the drug. However, there are also cases for which such risks are not binary. The information obtained from an experiment may enhance or diminish our expectations but not necessarily kill the drug. These are especially true for "product enhancement" programs by which a marketed drug goes through additional R&D to either improve the dosage form or expand the label (by adding more populations such as children or more indications). There are also drug interactions that may require further evaluation after the drug is on the market. It is impossible to test for every possible interaction (adverse effects if the drug is taken with other drugs) during the initial R&D process. Once on the market, the FDA may require additional studies to eliminate any such concerns. The company may also try to titrate the dose better (to achieve a more optimal dose based on population characteristics) by performing additional clinical trials after the drug reaches the market. Another possibility is conducting clinical trials in specific population groups (such as in Japan) who may require what is called "bridging studies" to take into account certain differences in how the drug is metabolized. In many of these cases, the company and the regulators are satisfied with the drug's safety and efficacy profile, and the additional experiments are either to expand its use or to eliminate certain concerns that may have surfaced post-marketing. Since we do not expect these experiments to fail outright, modeling private risks in a binary fashion may not be appropriate.

Consider an R&D program for product enhancement. For the first analysis, consider a program to improve the formulation of the drug. Currently, the drug needs to be taken once a day, and a new form is suggested that will allow the drug to be taken once every week. This "sustained-release" formulation requires R&D to devise a new manufacturing process. It then has to be tested in a population to ascertain the effectiveness. If it is found to be as effective as the once-a-day version, then both versions can be sold on the market. The once-a-week version may provide higher pricing power for the company due to the convenience it offers for patients and higher assurance of compliance it provides to doctors. A successful product enhancement of this type, then, will enhance the overall revenue for the company over and above what is expected currently from the marketed drug. The company considers the additional investment because it may enhance the value of the drug (because of the higher revenue).

Consider the product enhancement program shown in Figure 9.32. If the product enhancement goes as expected, we expect an increase in value of the drug by 20%. This is due to both a higher pricing power for the more convenient and compliant product and a halo effect it will create for the existing product. We value the current marketed product as $1.0 billion. As before,

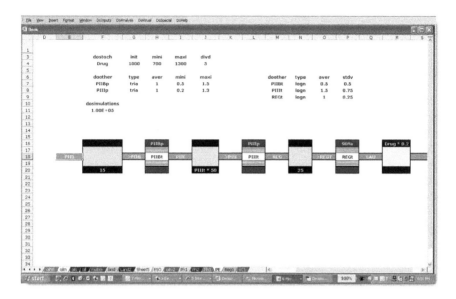

FIGURE 9.32
Product enhancement model.

this is the NPV of the free cash flows expected from the marketed drug (from today until the end of its life) using a discount rate commensurate with the systematic risk in these cash flows. This means that if we were to sell the drug to another company today, we should expect to get approximately $1.0 billion. The product enhancement program, including the regulatory process, is expected to take 3 years. By analyzing the uncertainties around product revenues, we also estimate a minimum and maximum of the value of the drug in the 3-year window. Such an analysis may also include any competitive actions and associated effect on the drug.

It requires a $15 million commitment up front to test the feasibility of creating a new form that will allow once-a-week dosing. It is expected to take 6 months (but this timeline is uncertain). We designate this as PIIBt. At the end of this experiment, the program will succeed with the probability, PIIBp. We expect this outcome to be 1.0 (the expectation we have going into the experiment). We can, however, be disappointed and get an outcome that is 0.5. In this case, our expectation of value from the product enhancement is reduced by 50% due to a technical problem. Or, we can get lucky, and the value enhances by 50% (a value of 1.5).

At the completion of this pilot study and after getting a technical reading that enhances or diminishes our expectation of value enhancement from the program, we have the option to abandon the program or invest more into it to manufacture the prototype at scale and test it. This is phase III. We expect this phase to take approximately 2 years (but it is uncertain, represented by PIIIt). The cost of entering this phase is a function of the time needed to complete the phase. The cost is $50 million per year (as it is an expense largely driven by people). If it takes 2 years to complete this phase, the investment needed will be $100 million ($50 million × 2). The scale-up and testing in PIII also carries significant risk. If everything goes as expected, the probability (PIIIp) takes a value of 1.0. However, this value can be anywhere between 0.2 and 1.3. The lower bound represents significant scale-up issues diminishing the value of the attempted product enhancement. Conversely, there is a possibility of a positive surprise as well.

At the completion of phase III, the data have to be submitted to the FDA for approval. It is going to take a total of $25 million in expenses and 1 year (the timing is uncertain and is represented by REGt). There is also a 90% chance that the FDA will approve the new formulation for sale. If so, the value of the marketed product is enhanced by 20% (represented by drug × 0.2). The gain from conducting this product enhancement program, then, is 20% of the value of the drug when the program completes.

This creates an additional complication. As the drug has a fixed patent life, its value will decrease as we progress in time as there is less patent time left. This is bit like a heavy dividend-paying stock. As described, as the stock pays a dividend, it loses value (as money is returned to investors).

In this R&D program, the underlying asset is the marketed drug, so the various investment choices are options on the value of the program. Figure 9.33 shows examples of the progression of value of the marketed drug. As can be seen in assumptions given further below, we model the value of the marketed drug with a dividend yield of 5%, so it is expected to lose 5% of its value every year.

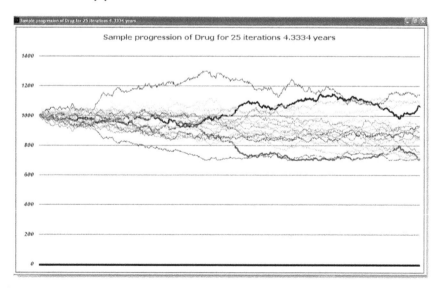

FIGURE 9.33
Stochastic simulation of the value of the marketed drug.

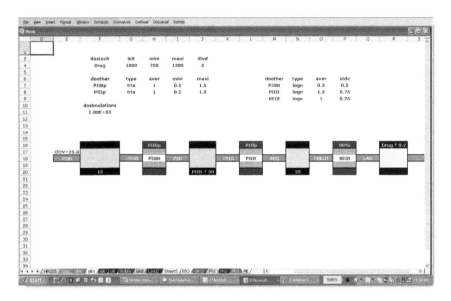

FIGURE 9.34
Value of the product enhancement program.

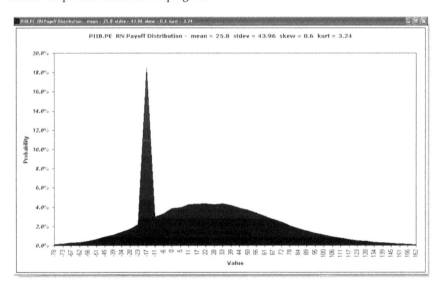

FIGURE 9.35
Risk-neutral payoff from the product enhancement program.

Using the assumptions described, we value the product enhancement program currently at $26 million as shown in Figure 9.34. The probability and cumulative probability of risk-neutral pay off from the investment are given in Figure 9.35 and Figure 9.36, respectively. The peak in the probability

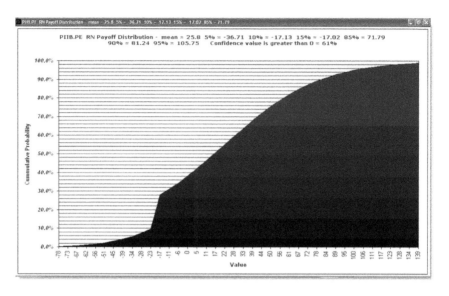

FIGURE 9.36
Cumulative probability of risk-neutral payoff from the product enhancement program.

distribution represents abandonment at the start of phase III. Since this problem does not involve a catastrophic technical failure, this peak represents a conscious decision by the manager of the program to abandon it. The cumulative probability of risk-neutral payoff illustrates that there is only a 60% chance of actually making an economic profit by entering the product enhancement program. For a company focusing on NCEs, such product enhancement programs are not necessarily attractive (either due to a higher cost structure or due to the opportunity cost of limited resources such as skilled personnel and space that could be utilized in more valuable R&D programs seeking NCEs). In this case, it may be better to attempt to out-license the product enhancement opportunity to another company with a lower cost structure.

This type of analysis is useful in both selecting higher-value programs that may be competing for limited resources (selection decision) and structuring out-licensing deals. For example, if the company decides to sell (out-license) the product enhancement program described here, it can negotiate a price close to $26 million. Since the buyer is also likely to perform similar analyses, it is unlikely that the transaction will happen at that price. If the company wants to retain at least 50% of the economic value in the product and yield the other 50% to the buyer, it may negotiate a price close to $13 million. More likely, the transaction will be a complex one that will include milestone payments (which happen at specific decision points) and royalty payments if the product makes it to market. Such a structure allows both the buyer and the seller to share the risk. We discuss such contract structures in more detail later in the book.

Regime Risk with Approval Alternatives in an R&D Program

In the previous cases, we learned different types of private risks (binary, continuous) as well as uncorrelated event risks incorporated in the price process of the underlying asset. In the case of event risks, we assumed that there is certain probability that an event may happen (anytime during the duration of the Decision Options problem) that instantaneously reduces the value of the asset by a certain percentage. In many cases, such events happen only in certain regimes. For example, we can be reasonably sure that if no unknown product risk surfaced in the first 3 years of the program, it is less likely to occur later. Similarly, the risk of competitive entry may be higher a few years after launch as competition gathers data on the marketed product and reengineers/redesigns the product to be more effective, less toxic, or both. Competitive entry probability is also high during the year of launch as competitors rush to market to capture any benefits of being the first entrant. Such efforts take time for R&D, so competitive entry is more likely to occur the first couple of years after launch. It may also be a strategic decision by the competitor to wait to see how the product is accepted and formulate a marketing strategy for its own product that could be more effective. Whatever the reason, the probability of certain events to happen is higher in certain times of the drug's life.

Consider the R&D program shown in Figure 9.37. The drug candidate is currently completing phase IIA, and a decision has to be made to start phase IIB in 6 months. If phase IIB is kicked off, a commitment of $25 million is needed. PIIB is expected to last 2 years and has a success rate of 70% (binary).

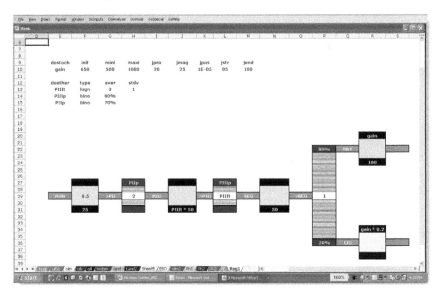

FIGURE 9.37
R&D model with approval alternatives.

If the drug succeeds PIIB, then it can enter phase III, which typically lasts 3 years (but it is uncertain with lognormal characteristics and a standard deviation of 1 year). The cost for phase III scales with the total duration of the phase and is expected to be $50 million/year. There is an 80% chance of success (binary) from phase III. If the drug succeeds in phase III, an NDA (New Drug Application) can be filed at a cost of $25 million. The FDA decision is expected 1 year after filing. However, the approval is not as straightforward as previous cases. The drug is expected to be approved, but there are two possible outcomes; the first one is a regular approval, in which case the drug can be launched at a cost of $100 million. This is the likely outcome (80% chance). But, there is a 20% chance that the FDA will require a "black box" label on the drug—a strict warning regarding side effects and constraints on how the drug can be prescribed and administered. If this is the case, the value of the drug will be only 10% of the expected value under regular approval. The company will sell the drug at that stage to another firm with lower cost structure or more specialized expertise in marketing the drug with black box labels for the disease targeted.

To compute the economic value of this R&D program, we have to estimate the value of the drug at launch (under regular approval). Using market research data and applying regular DCF analysis using a discount rate commensurate with the systematic risk in the free cash flows from the drug, the following assumptions are reached. The average expectation of value at launch is $650 million with a range of $500 million to $1 billion. This excludes the launch cost.

There is one other complication. In the last year of the R&D program, there is a 20% chance of a competitor entering the market. While the drug waits for FDA approval, the competitor's product can garner valuable market share, leading to a 25% loss in the value of the drug. We represent the stochastic process of the drug's value in the fashion shown in Figure 9.38.

The value of the drug (called *gain*) is in the range $500 million to $1 billion with an expectation of $650 million at launch. There is a 20% chance a "jump" happens (jpro, representing competitive entry), and if that happens, the magnitude of the jump is 25% (jmag, representing loss of value). All jumps are negative (value is always lost), and the percentage of positive jumps (jpos) is zero. The jumps start 85% into the R&D program (from now). Since the total R&D duration to launch is 6.5 years, this means that jumps start 5.5 years into the R&D process. Note that since the total duration is uncertain, the precise start of jumps is not known. The "regime" in which jumps occur starts 85% into the R&D process and ends at launch (or licensing out in the case of the black box label).

Figure 9.39 shows examples of the progression of the value of the drug with these characteristics. The arrow indicates a scenario in which a competitive entry has occurred and has resulted in a corresponding loss in value. Also shown is the regime during which jumps are assumed to happen. Figure 9.40 shows the risk-neutral payoff from the analysis. The R&D

dostoch	init	mini	maxi	jpro	jmag	jpos	jstr	jend
gain	650	500	1000	20	25	1E-05	85	100

FIGURE 9.38
Description of stochastic process of gain.

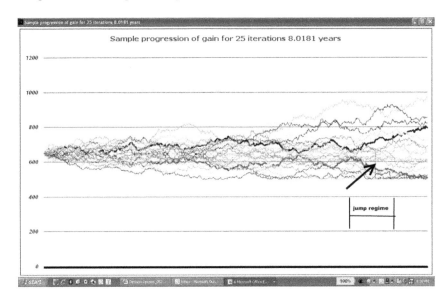

FIGURE 9.39
Stochastic simulation of gain with jump regime.

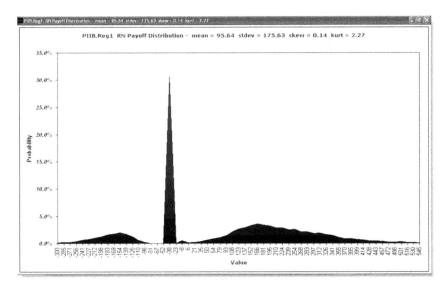

FIGURE 9.40
Risk-neutral payoff from the R&D program.

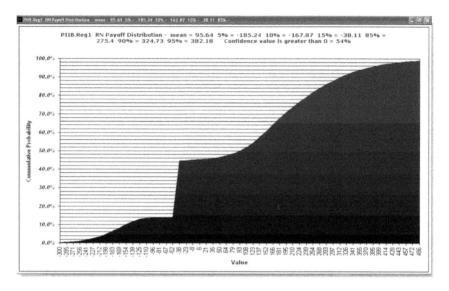

FIGURE 9.41
Cumulative probability of risk-neutral payoff from the R&D program.

program is currently valued at $95 million. The peak represents either technical failure or abandonment after phase IIB. The negative tail is due to technical failures after phase III and situations in which a black box label was required and the drug's value is low.

Figure 9.41 indicates the cumulative probability of risk-neutral payoff indicating close to a one-in-two chance of making an economic profit from the program. The regime risk (risk that happens only within a window of time) may be due to a variety of reasons. In this example, we considered competitive entry as a regime risk. Regulatory actions (which are uncorrelated with the market) may also be modeled as a jump in the stochastic process that models asset value. Such a risk is higher after the conclusion of an experiment and the filing of regulatory materials. The economic value of the program depends on when decisions are taken for further investment (whether such decisions are made before or after the resolution of the regulatory risk), although project managers intuitively know this, and they add flexibility into the design of the experiment to stop or scale down if an unfavorable regulatory action materializes. Because of the different influence on value by a variety of such factors, a holistic analysis is always better for making a project design and selection decision.

Portfolio Diversification in Large Companies

The established notion in finance is that a diversification discount exists for conglomerates with disparate businesses because it is easier for investors to diversify than it is for companies to do so. Investors like pure plays, and each investor will form a portfolio that is on the efficient frontier by assembling a well-diversified group of companies on his or her own. We can take it one step further and hypothesize that the investors also prefer one-product companies (the ultimate pure play) compared to a company that has a number of similar products. If this is the case, investors will prefer one-product biotechnology companies to pharmaceutical companies as they can buy a number of biotechnology stocks to replicate the portfolio of a larger pharmaceutical company. For this to be true, we also have to assume that there are no scale effects in larger companies, and they do not gain any advantages in transactions with external parties such as suppliers and investigators.

There are indications that scale and transaction-based advantages have indeed declined for large companies. However, there may be capital-related advantages for a large portfolio company in life sciences. Early withdrawal of capital (by venture capital firms or other entities that fund smaller companies) can result in a loss of value for biotechnology companies. Small portfolio companies also face high private risks (undiversified private risks emanating from few products), resulting in high volatility in their equity that may exclude them from certain public funding sources (such as mutual funds and university endowments that may impose artificial constraints on single-position volatility even though such volatility is uncorrelated with the market), creating a market failure. In such situations, the theoretical arguments that hold in perfect markets around cheaper diversification by investors may not be valid.

Let us visualize how portfolio diversification changes the payoff expectations in small and large companies. Consider the R&D program from the preceding case. Assume that this is the only program under development in a small company. The expected risk-neutral payoff (which is close to the actual payoff as the correlation with the market is low) is shown in Figure 9.42. This looks very risky indeed as the company may lose as much as $300 million in the worst case. In any scenario that is less than zero, the company is effectively bankrupt as the venture capital firms that fund such a company will likely pull all promised funding. Now, consider a large company that has 50 such programs under development. The risk-neutral payoff of this large portfolio is shown in Figure 9.43.

The first thing to notice in Figure 9.43 is the difference in payoff expectation. This is the average payoff, so the total payoff from the portfolio is 50 times the value here as the larger company has 50 similar programs in R&D. The range of average payoff from a single program is from $30 million to $150 million. (Remember that the single company payoff ranged from −$300 million to $500 million.) So, for a portfolio of 50 such candidates, we can expect an economic value of $1.5 billion to $7.5 billion, and the chance of a loss in economic value is virtually eliminated.

Figure 9.44 shows the cumulative probability of risk-neutral payoff. Notice that there are no negative outcomes in the average payoff, and we are nearly 100% confident that the portfolio has a positive payoff.

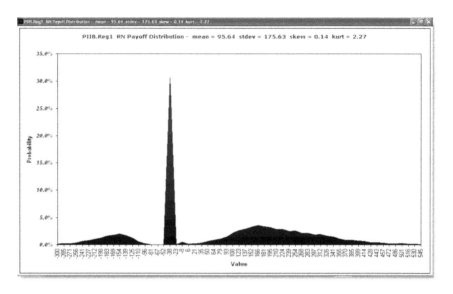

FIGURE 9.42
Risk-neutral payoff from a single R&D program or a company with a single program.

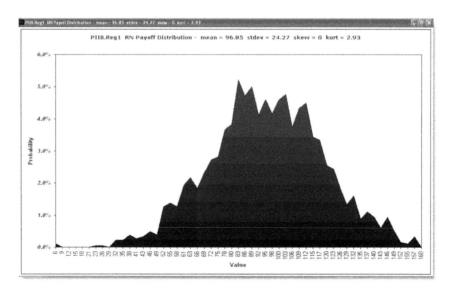

FIGURE 9.43
Risk-neutral payoff of a diversified company with 50 different R&D programs of similar types.

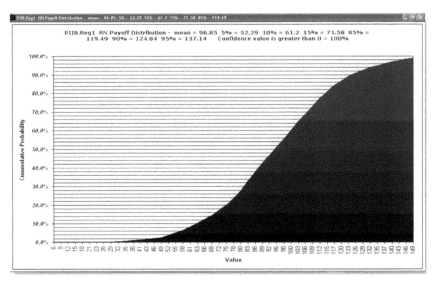

FIGURE 9.44
Cumulative probability of risk-neutral payoff of a diversified company with 50 different R&D programs of similar types.

The question is whether there is a premium (or discount) for the large company performing the portfolio diversification as opposed to an individual investor (by buying a large number of small single-product companies). The theoretical argument here is that an individual investor can assemble a similar portfolio as held by the pharmaceutical company by buying 50 single-product companies, and so the investor can manufacture the portfolio return of the pharmaceutical company at a cost less than what it takes a pharmaceutical company to do so. Unfortunately, this is not simply a theoretical finance question as many other factors such as learning from repeated experiments in a large company, market failures due to short windows of capital providers in small companies, meddling in operations of smaller companies by capital providers, lack of resources to assemble the right skill set in smaller companies, and many other factors point to a more efficient portfolio diversification mechanism in a large company.

The ability to diversify private risk in a portfolio of uncorrelated projects is clear. However, it is unclear who is in a better position to diversify: an operating company (such as a large pharmaceutical company) or an individual investor buying stock in a large number of diverse single-product companies. The real asset diversification conducted by the pharmaceutical company is technically equivalent to the financial diversification attempted by the investor. The trade-off is the scale and learning advantages in the pharmaceutical company against the cheaper and liquid diversification benefits in a financial portfolio. Further research is needed to identify actual benefits and costs and determine whether investors place a premium (or discount) for the diversification provided by the pharmaceutical companies.

Rigid Budgeting and Capital Constraints Leading to Loss of Value

Budgets and other artificial capital constraints can have a significant effect on the value of R&D programs. The budgeting process in a large company can get complex because of the large number of different departments working together on a large number of R&D programs in different stages of development. Because the business is both people and information intensive, the budgeting process needs to include different resources such as people/skills, money, time, equipment, and space. Generally, all components are driven down to the dollar using conversion rates that can hide specific resource constraints that can affect the progression of R&D or, in some cases, premature termination solely due to lack of a specific type of resource. Small, one-product companies are generally supported by venture capital firms, at least partly. The venture capitalist generally has a portfolio of small companies, in effect replicating the portfolio of a larger firm. Although conceptually it is clear how an individual investor can create a fully diversified portfolio without cost in a public market, the advantages of a venture capital firm doing so is unclear. It is not the diversification benefits that motivate the venture capitalist but market inefficiencies such as information inefficiencies related to privately held firms in which they typically play. However, once having assembled a portfolio of privately held firms, the venture capitalist may act similar to managers of a larger firm (with an internal private portfolio). In these situations, if the venture capitalist does not have a good understanding of the operating characteristics of individual companies, technologies, and products, his or her involvement in operating decisions can destroy value. The trade-off is the information advantage gained by the venture capitalist in identifying and investing in private companies against the operating disadvantage of a complex portfolio of not only products but also companies. In this sense, venture capital firms that focus on the financial aspects of their portfolio take a more hands-off approach to operating aspects and may have a higher chance of enhancing shareholder value for their investors.

Consider the R&D program shown in Figure 9.45. Assume that this is the only program in a small company. If this program is funded, we calculate the value of the program (and the company in this case) to be $26 million.

The budgeting process (using large spreadsheets that take into account all types of costs) may have concluded that the company needs a total of $265 million to take the program to completion. From the assumptions in Figure 9.45, it can be seen that $15 million is needed in stage I, $100 million in stage IIA, $100 million in stage IIB, and $50 million in stage III. These are the average expectations of the total costs in the company. Suppose we fund this company with $265 million as a hard constraint. That is, the company gets only $265 million, and if it cannot make a transaction (i.e.,

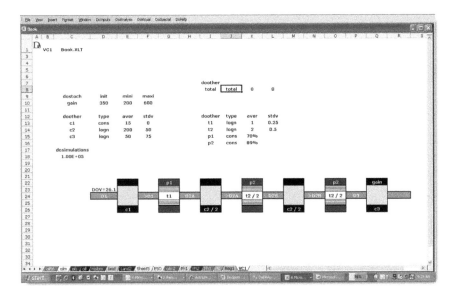

FIGURE 9.45
Multistage R&D program.

get the drug to market, sell it, or issue an initial public offering [IPO] for further capital) within that cost, no more money is provided by the venture capitalist. This is a hard capital constraint as the original capital provider owns a substantial part of the company, and if he or she does not choose to participate in the next round (assuming that it is needed), the chance of acquiring additional capital by the company (over and above the original $265 million) is very slim. Life sciences IP is also complex and requires significant work to sell to another party (assuming a buyer can be found) prior to the completion of a critical experiment that the company may already have entered. That is, if the company runs out of money in the middle of an experiment and the original capital provider does not provide additional resources, it is pretty much stuck—unable to acquire new capital or sell the existing IP to another firm.

The company can be valued by applying a hard constraint of $265 million. In this case, we value the R&D program (the only IP position within the company), assuming that the program will be abandoned and the company will cease operations if it runs out of the initial capital provided ($265 million). Figure 9.46 and Figure 9.47 show the results and the risk-neutral payoff, respectively, from such a company that faces a hard capital constraint.

The peak in Figure 9.47 represents abandonment or technical failure after stage I. In this case, the company folds after the first $15 million, and the capital provider loses the $15 million and walks away. The rest of the losses to the left are related to technical failures, abandonment, or simply running out of money. As can be seen, imposing an *ex ante* hard capital constraint

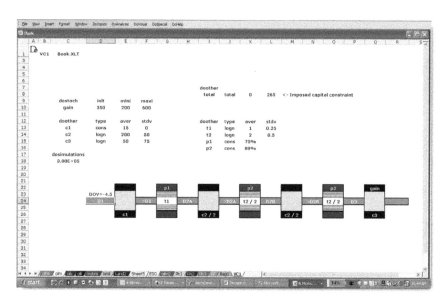

FIGURE 9.46
R&D program funded with a rigid capital constraint.

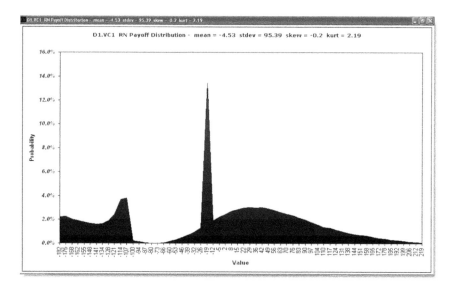

FIGURE 9.47
Risk-neutral payoff from an R&D program with a rigid capital constraint.

diminishes the value of the company from the original $26 million to −$5 million. This means that investing $265 million (only) into this company (with a single R&D program) is a negative NPV project. It does not make sense. The capital provider has to make decisions at a later time considering all available information. In life sciences, it is difficult to get new information on a drug's prospect when a company is in the middle of an experiment.

Clinical trials are typically double blinded. The trial is conducted with a control group. Some part of the patient population receives the drug, and the remaining group receives a placebo (a sugar pill that is designed to look like the drug tablet). *Double blinded* means that neither the investigator (clinician conducting the trial) nor the patient knows who is getting the drug and who is getting the placebo. Once the trial is completed and all the data are available from the study, the data can be analyzed, and the results from the population who received the drug can be compared to that of those who received the placebo (control group). If the company ran out of money during the trial, it may have to suspend the trial, complete its regulatory obligations, and walk away. More likely, the company may have completed one arm of the trial and exhausted its resources. Before a conclusion can be reached, the situation may require additional studies. Whatever the case, assume that the hard capital constraint is in effect. If the company runs out of the original "budget," it has to stop everything.

Why would anyone impose such a constraint? There are many possible reasons. The capital provider may use the constraint as a "disincentive" on management so management will be more resource conscious. The managers know the implications of running out of money up front. Another reason is downside risk management in the portfolio. The capital provider may have invested in many similar companies and wants to control the risk of overinvestment in any specific company. It may also be "irrational risk aversion"— the investor's risk tolerance decreases after the original investment and he or she starts to consider the "sunk costs" in the decision-making processes. This is dangerous for the capital provider. The provider becomes regretful of the sunk costs and walks away from a valuable asset because the uncertainty has not completely resolved and the traditional financial methods are not capable of assessing economic value.

This type of capital constraint can also exist in large companies, although it is less likely. Decision makers may use proxies such as sunk cost as a measure of attractiveness of the program. If the program is more costly than anticipated or slow to progress, decisions may be based on only the costs (without consideration of the overall value). Since costs are more "visible" in a long-drawn R&D process in life sciences, unfortunately costs sometimes do become the only proxy for decisions. The abandoned R&D programs and associated IP also are locked up in most large pharmaceutical companies as a sale of these programs to other companies (with differing cost structure, scope, and specialization) is not typically pursued as an alternative for a variety of reasons, including the fear of losing IP from active programs.

Another type of capital constraint is a hard yearly budget. In this case, the capital provider sets a hard yearly budget, and if the budget is exceeded, refuses to fund the program (or company) further. This may be another

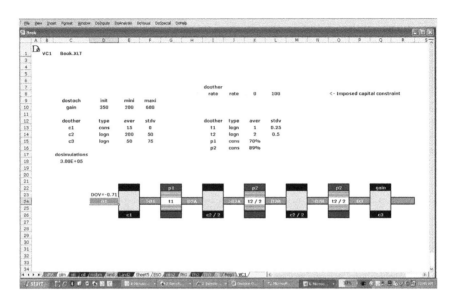

FIGURE 9.48
R&D program with rigid yearly budget.

situation in which a capital provider may be using the yearly budget as a disincentive against underperformance or overexpenditure. Assume that in the R&D program and the one-product company discussed the capital provider has set a constraint of $100 million per year. This means that if the project manager runs out of the yearly budget, the manager is forced to abandon the project. The $100 million is reached arbitrarily. In this case, it overfunds the program early and underfunds it later—almost the exact opposite of what may be optimal for the company from a value perspective. Figure 9.48 shows the value of the program and the company under this constraint. Again, the values of the program and the company are reduced to close to zero. A hard yearly budget such as this (we are assuming that once the budget runs out, we simply abandon the program) can also happen in a large company.

Certain companies continue to spend large amounts of money and time in precise projections of quarterly EPS, guidance setting, and meeting of the EPS. It is true that if the company precisely forecasts an EPS a few weeks before it is released and fails to meet the forecasted EPS, its stock will take a tumble. It is not necessarily because the market is "disappointed" at not getting this quarter's EPS, but rather it is disappointed at the management's inability to forecast and manage the company. If the managers of the company "miss" such a tactical measure (which affords significant "accounting flexibility"), the market may lose confidence in the management. In information-driven industries that show high levels of uncertainty such as life sciences, energy, and technology, it is nearly impossible to project profits precisely.

There are a few companies that have understood this concept and have decided not to forecast EPS or to give any guidance to the market. They have decided to give the market the uncertainties involved in its business and the primary parameters that drive the company's profitability. This is much more relevant information for the market in projecting the company's value compared to a precise (and deterministic) forecast of EPS. There are a number of "value-destroying" actions that a company could take to attempt to meet the tactical EPS (such as shutting down R&D programs, delaying critical purchases, hiring freezes, termination of leases, etc.) once the EPS projection is "promised" to the market. By avoiding such actions and focusing the company's management on value-increasing actions, companies can substantially improve their long-term strategic position. It can also be noted that projections of deterministic EPS estimates reduce management flexibility, and that alone can reduce the value of the company.

There are number of examples of companies and industries that have lost their dominant positions because of a focus on tactical and deterministic financial measures. Some of this is due to experiences and training from the production economy (where quantity and cost were supreme) applied in a dramatically different world of the information economy that is driven by uncertainty and intangible assets such as IP. Assuming that such companies could be run by tactical financial measures is a gross misunderstanding of how value is created in this environment. It is the ability of the management to create, nourish, and optimally exercise a portfolio of options that is critical in shareholder value improvement. Good managers get paid handsomely not because they are great forecasters of the future but because they have an innate ability to manage uncertainty. There is no avoiding the uncertainty by precise forecasting. Devising systems, processes, and methodologies to take advantage of uncertainty differentiates management. It is for this ability that they earn their keep.

It is difficult to replicate good management intuition and decision making under uncertainty. However, it is possible to approach these situations more systematically if managers own up to the existence of uncertainty and proactively manage it using analytics similar to those cited in this book. These types of analyses do not replace the need for good management intuition but aid in its development and nourishment.

Structuring and Valuing an External Transaction

Consider a pharmaceutical company Maxo that has been active in the area of cancer research for a number of years. It was a strategic shift for the company when it decided to ramp up its oncology (cancer-related) R&D. In doing so, it also invested in sales and marketing of oncology drugs in anticipation of new drugs coming out of its newly focused R&D. As luck would have it, the R&D machine has not kicked into gear yet, and the company is in need of new products to keep its marketing machine busy and its revenue line growing. One way to do this is to "buy" drug candidates in development from other companies.

One such company is Egen, an up-and-coming biotechnology company formed 5 years ago by a few budding entrepreneurs, scientists, and academics. Egen has been focusing on the discovery of oncology drugs for all types of indications. It has been firing on all cylinders and now has three drug candidates in advanced discovery and toxicology testing stages. One such candidate, Edoxin, is getting ready to enter the clinic for the first experiments in humans. Egen has already completed toxicology studies in animal models such as rats and dogs, and everything looks good thus far.

On a fine fall afternoon as the Egen managers were conferencing on the development plan on Edoxin, they got a surprise call from the chief scientific officer of Maxo. Maxo has been studying Edoxin after reading an article published by one of Egen's enterprising scientists. It looked very interesting to Maxo, and the mechanism of action is something they are pursuing as well. Maxo has a number of discovery programs in this area but none as advanced as Edoxin. Maxo needs a drug soon to rev up its revenue growth, and Edoxin appears to fill the need.

For Egen senior management, this was a welcome call. As anybody who has gone through a start-up knows, cash is king. They had venture capital funding when the company started. Further rounds of financing got them to the current stage, but they are fast running out of money. Everybody knows they have valuable assets in the pipeline, but unless they can be "monetized," it does not help their cash flow situation. The venture capital firms are always interested in "exit strategies," and some of the early investors are keen on creating internal cash flows to fund newer discovery programs rather than investing more funds.

There may be a motivated buyer in Maxo and motivated seller in Egen. Let us take a closer look at Edoxin (Figure 9.49). Egen has big plans for Edoxin and an elaborate development plan. First, they plan to test the drug for a relatively small indication, pancreatic cancer. This is the first indication for which the drug has shown efficacy in animal models. But, they do not plan to stop there. If it succeeds in treating pancreatic cancer, the company plans to try the drug for various other indications in the cancer area, specifically lung, neck, and breast cancer. Smaller companies like Egen economize on development programs by staging them, in this case trying one indication first and expanding it to other indications once the lead indication succeeds and is approved. There are advantages and disadvantages to this strategy. An advantage obviously is that they commit small investments early, prove the prototype, and then expand (substantially reducing the downside risk and increasing decision flexibility by delaying decisions). The decrease in downside risk, however, comes at a cost of reduction in upside potential. The first loss of upside potential comes from the delay in getting to market and corresponding loss in usable patent life after approval. Remember that the composition of matter patent on the

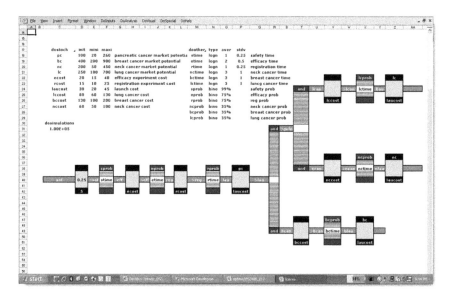

FIGURE 9.49
R&D project plan for Edoxin.

chemical will be taken during the trials of the pancreatic cancer indication. By delaying the testing and possible marketing of "larger" indications, the company will have less time on the market for those indications (provided of course it succeeds and gains approval). The other disadvantage is that the delay gives the competition a leg up. Once the drug is on the market, competition can study it as well as all the trial data. This may provide them with ammunition to accelerate their own programs, possibly for indications targeted by the company's product.

All these are true, but for a small biotechnology company in a current cash-strapped situation, the best it could do is to push forward with the pancreatic cancer indication and hope for the best. If the drug does get to market, Egen holds valuable expansion options it can exercise or sell. These can be exercised if Egen is successful in raising more money as the purse strings of the venture capitalist may be a little less tight as he or she watches positive cash flows from the approved indication and possibly the widening eyes of the big pharmaceutical companies thinking of acquisitions.

Egen has done an initial analysis of Edoxin's development plan to ascertain the economic value. To do this, it has collected the following types of data:

1. Market size, patients seeking treatment, pricing power, and reimbursement patterns to define the revenue growth. In doing so, Egen has also considered the possibility of competitive entry, regulatory and consumer attitudes toward cancer treatments, as well as the approaching presidential elections and saber rattling by certain politicians against pharmaceuticals.

2. Product timelines based on industry norms and benchmarking data. It has not done any development programs.

3. Cost estimates from a CMO to which it has outsourced the manufacturing because it has very limited manufacturing facilities to produce the chemical at scale. They also provided them a manufacturing plan and associated uncertainties such as availability of raw materials, lead times, yield, and other aspects.

4. Design of clinical trials, including the number of patients and dose arms needed based on the hypothesis being tested. In this context, they have also gathered estimates from a CRO Egen selected to conduct the clinical trials. The CRO has also provided estimates of clinical investigative sites and expected enrollment rates of patients.

5. Based on information gathered from their suppliers, the CMO, and CRO, Egen has modified its project timelines and uncertainties, and also calculated overall costs and uncertainties in various phases.

6. Egen also gathered industry benchmarking data on success rates in various stages for the indications pursued. Its experts then applied some adjustments to the success rates based on the data from the preclinical experiments and discovery chemistry.

Based on these data, Egen compiled the assumptions shown in Figure 9.50.

First, it used standard Monte Carlo simulation to estimate the value at launch of the drug for the various indications. For example, for pancreatic cancer, it made the following assumptions:

Peak sales	Average of $50 million and a standard deviation of $20 million
Net margin	Average of 40% (range of 30% to 60%)
Real discount rate	Risk-free rate of 0 (as the cash flows are not correlated with market). This is because diseases such as cancer need therapies whether the economy is expanding or contracting (or markets are going up or not). This is not the case for lifestyle and other types of drugs that are not lifesaving.
Time to reach peak sales	3 years
Time of stable sales (peak)	1 year
Time of declining sales	4 years
Terminal value	$5 million

Egen also assumed that the drug revenue growth would be smooth from 0 to the peak sales in 3 years. After the peak sales are reached, revenue drops to close to zero, smoothly over 4 years. It also assumed a terminal value of $5 million as the drug may still stay in the market for a while. Using the Monte Carlo functionality of the Decision Options software, it simulated these uncertain cash flows after approval and calculated the NPV (Figure 9.51).

dostoch	init	mini	maxi		doother	type	aver	stdv	
pc	100	20	260	pancreatic cancer market potential	stime	logn	1	0.25	safety time
bc	400	200	900	breast cancer market potential	etime	logn	2	0.5	efficacy time
nc	200	50	450	neck cancer market potential	rtime	logn	1	0.25	registration time
lc	250	100	700	lung cancer market potential	nctime	logn	3	1	neck cancer time
ecost	20	15	40	efficacy experiment cost	bctime	logn	3	1	breast cancer time
rcost	15	10	25	registration experiment cost	lctime	logn	3	1	lung cancer time
laucost	30	20	45	launch cost	sprob	bino	99%		safety prob
lccost	80	60	130	lung cancer cost	eprob	bino	75%		efficacy prob
bccost	130	100	200	breast cancer cost	rprob	bino	75%		reg prob
nccost	60	50	100	neck cancer cost	ncprob	bino	35%		neck cancer prob
					bcprob	bino	35%		breast cancer prob
					lcprob	bino	35%		lung cancer prob

FIGURE 9.50
Assumptions in the R&D program.

FIGURE 9.51
Monte Carlo parameters for NPV of the marketed drug.

Egen introduced two types of uncertainties. (1) It assumed that the peak sales were lognormally distributed with an expectation of $50 million and a standard deviation of $20 million. (2) It assumed that the net margin (after cost of goods and general marketing and administrative costs) was 40% and could be in the range of 30% to 60%. Ignoring all taxes, it ran a Monte Carlo simulation on the NPV of the free cash flow and obtained the following results. The value of the pancreatic cancer indication at launch was $100 million (with a range of $20 million to $260 million). Figure 9.52 shows the distribution of NPV from the simulation. Egen used a real risk rate of 0 for the NPV simulation as well.

It also conducted similar simulations on other indications, but this time reducing the expected time on the market as the product for these indications will reach the market approximately 3 years after the launch of the

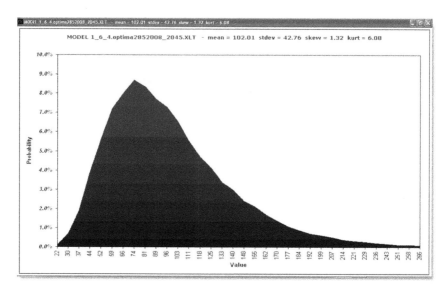

FIGURE 9.52
Probability distribution of NPV for the first marketed indication.

drug for pancreatic cancer. It also decided to model costs and timelines using lognormal distributions (taking into account all uncertainties). It also considered the success rates as binary (it was a go/no-go type decision).

On the cost side, Egen modeled the costs in different stages of pancreatic cancer (there are three stages: safety, efficacy, and registration) as independent but stochastic. This means that we can observe costs over time and get better estimates as they go forward in the R&D program. Once the drug is on the market for pancreatic cancer, Egen can consider the expansion options for the other three indications. These indications do not need safety studies as the safety has already been proven. They do need efficacy studies or another supplemental filing with the FDA. Egen shrank these activities into a single phase as the chances of not filing are negligible if the efficacy study succeeds, and the chance of approval is high. It combined the costs and probabilities into a single phase for the expansion indications. In all cases, it calculated an overall cost (with uncertainties) and an overall timeline (using industry benchmarks and feedback from the engaged CMO and CRO).

Oncology programs such as this one have generally high success rates. One reason is that they are treating diseases that do not have cures, so the objective is extending life. Toxic effects and TR have to be evaluated differently. In this case, Egen assumed a very high success rate from the safety experiment (99%), but the efficacy success rate was a more conventional 75%. For expansion indications, it combined the efficacy and approval probabilities into one and assumed that each indication had a success rate of 35%. This is lower than what was assumed for the lead indication as Egen had less confidence that the drug would be effective in these expanded indications.

One of the questions that came up during the modeling session (which was done in a single afternoon) is the correctness of modeling the value of indications and R&D costs as stochastic (following GBM in a risk-neutral framework). Some in the room said that they could see how the value of indications can have a proxy in the market, but they struggled with considering the costs in that fashion. They however concluded that the systematic risk of these cash flows (for a cancer drug) was low, and even for the traditional CAPM-based DCF analysis, the appropriate discount rate was the risk-free rate. Also, all revenues and costs were not inflation adjusted, so the appropriate discount rate was the real risk-free rate. Since the duration of the problems was approximately 6 years, a 5-year T-bond yield may be appropriate. Given the low yield (3%) and generally high inflation (close to 3%), Egen decided to use a real risk-free rate of 0. It also decided not to worry about taxes until it set out to calculate the value of the R&D program.

Given all the assumptions and a rather complex-looking model, the first thing Egen did was a traditional DCF analysis. Figure 9.53 shows the results from the DCF analysis. The results were a bit discouraging. The lead program was worth less than $1 million. This did not seem correct so participants rechecked all inputs but found them to be correct. Someone suggested that perhaps the discount rate was causing the low NPV, but everybody quickly realized that it could not be any lower than the 0% real discount rate that they assumed. Looking through the results, they found that there were negative NPVs emanating from the three expansion arms of the decision tree. This also seemed counterintuitive as Egen did not necessarily have to invest

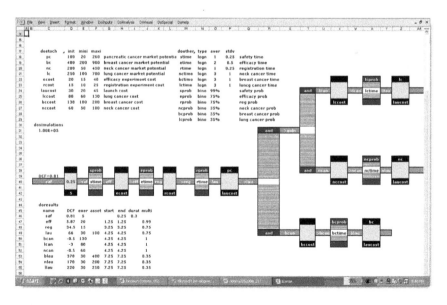

FIGURE 9.53
DCF NPV of the entire R&D program.

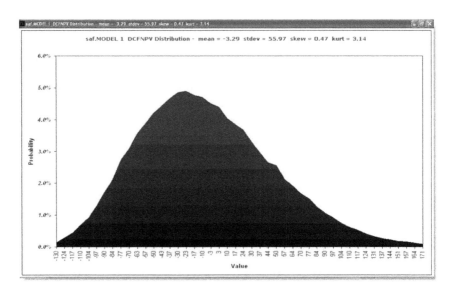

FIGURE 9.54
Monte Carlo simulation of NPV for the R&D program.

in these if they were of negative value to the company. Someone suggested that perhaps they should remove the expansion indications, which would bring the NPV close to $5 million. Removing the options to expand the drug improved the NPV—again a counterintuitive result.

The next step was to run a Monte Carlo simulation of NPV. There was so much uncertainty in all the inputs that taking only the average value (as is done in traditional DCF analysis) may not have been proper. The result from the Monte Carlo simulation of NPV is shown in Figure 9.54. Egen modeled the success rates as average probabilities. This is similar to running a scenario analysis such as optimistic, average, and pessimistic. In creating the NPV distribution, Egen ran 100,000 different scenarios with the cash flows adjusted with the average probability of success rates. Although the expected value of the NPV was close to zero, some felt a little better with this result, which took into account all the uncertainties they knew existed. Some really focused on the right side of the distribution and wondered if they could have more positive values and avoid the negative ones on the left would improve the result. It was clear that the program could indeed be very valuable, although on average it was not.

After some discussions, they decided to run a Decision Options analysis, this time allowing the decisions to be options. This means that each decision to commit investment into a stage was an option. It was a right but not an obligation. The same held true for the expansion options. After the pancreatic cancer approval, Egen would have had better information on the costs of running other experiments for newer indications as well as the market potential of new indications. If the potential was high or

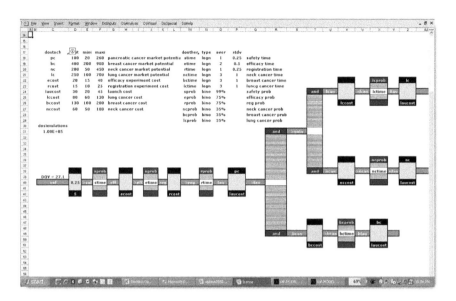

FIGURE 9.55
Completed analysis of the R&D program.

costs low, Egen would have the right to try them, essentially capturing the right side of the probability distribution of NPV. But, if potential was low or costs were high, the company simply could not do it, thus avoiding the negative side of the distribution. What Egen wanted (capturing more of the positive outcomes and avoiding more of the negative outcomes) was in fact possible. It was intrinsic in the design itself. These decisions were true options. Considering such decision flexibility (inherent in options), combined with all the uncertainty, Egen conducted a complete Decision Options analysis (Figure 9.55). Figure 9.56 shows a better result. The value of the program was approximately $27 million if the flexibility in the decisions was considered. In this case, Egen also considered the success rates to be binary.

There are two important aspects to note from these results. First, the range of risk-neutral payoff was from −$300 million to $800 million with an expectation of $27 million. The low values occurred in undesirable scenarios when the expansion option investments were all taken, and they either technically failed or resulted in very low-value products so that it is better to walk away rather than take them to market.

Figure 9.57 shows the cumulative probability of the risk-neutral payoff from the program. The cumulative probability distribution shows that that the program has only one in four chances to make money. However, the positive skew of the payoff makes it valuable enough to undertake. It is also the case that there is a 15% chance that the program will lose more than $65 million but an equal chance that it will make over $110 million.

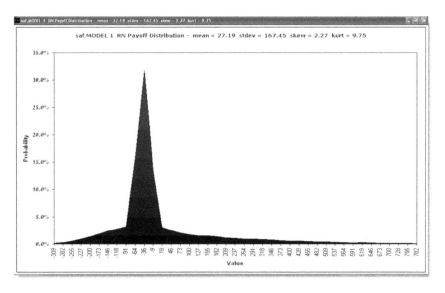

FIGURE 9.56
Risk-neutral payoff of the R&D program.

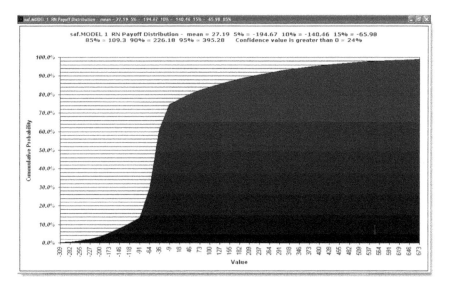

FIGURE 9.57
Cumulative risk-neutral payoff of the R&D program.

The analysis gave the Egen team a reserve price for its program. They now knew that the intrinsic value of the program was approximately $27 million. If Maxo were to buy the program outright, Egen should at least get $27 million. Egen also felt that since it was unlikely that Maxo had done an analysis similar to this one, they should start at a higher number and negotiate downward if necessary. In this case, Egen decided to ask for $50 million for the transaction. That would be sufficient for it to pursue five new discovery programs, and this model of bringing candidates to the door of the clinic and selling them may be a better model than undertaking an extremely risky and expensive development program. A large pharmaceutical company like Maxo is better able to diversify such risk by pursuing a portfolio of similar programs.

So, Egen's chief executive officer (CEO) contacted Maxo and discussed what Egen believed the value of the program to be. After a few days, Maxo came back with a different type of deal. Maxo said that it was not interested in buying the program outright but would like to enter into a licensing deal with Egen. Maxo understood that Egen was in need of some immediate cash, and was willing to pay $3 million up front. The rest would come in milestones and royalty payments. Maxo suggested the following licensing structure:

Up-front payment at contract signing	$3 million
Milestone payment at NDA filing	$5 million
Royalty on net income (on all marketed drugs)	10%
Milestone payments at the start of expansion indications	$3 million each

The milestones and royalties suggested by Maxo appeared low to Egen. The first thing Egen did was to go back to its valuation model and revalue it with the milestone and royalty information. Since it established that the value of the current program was $27 million, it wanted to find the value that Maxo would get if Egen accepted the suggested licensing structure. The result is shown in Figure 9.58. Egen included the $3 million in contract signing and the $5 million milestone payment at NDA filing (REG phase) as additional costs to Maxo. Remember that if Egen signed off on the deal, Maxo would take over the program and would then be responsible for all R&D costs. Maxo would pay Egen $3 million up front. This is an additional cost for Maxo, over and above what it would cost it to run the safety experiment. If the drug candidate reached the registration phase, Maxo also would be responsible for all the costs, and would have to pay an additional $5 million in milestone payment to Egen. If the drug was launched, Maxo would also have to pay 10% royalty on net income. This in effect reduced the value Maxo would get from the marketed drug by 10%. Maxo represented this in the launch phase by reducing the NPV by 10%. Also, note that there were additional milestones to be paid ($3 million each) if Maxo decided to progress the drug into expansion indications. These were additional costs for Maxo, and it represented them appropriately. The royalty reached through to additional indications as well. Egen had a 10% claim on any indication approved and marketed by Maxo. This means that the value of all marketed indications by Maxo had to be reduced by 10%.

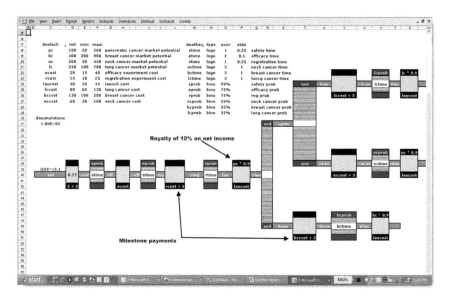

FIGURE 9.58
Analysis of a licensing deal for the R&D program.

After incorporating the components of the suggested licensing structure, Egen management revalued the R&D program from Maxo's perspective and found that the value was approximately $15 million. This means that Maxo gained $15 million in shareholder value from the R&D program. This also means that Egen's gain of value was $12 million (the original $27 million minus $15 million). The suggested licensing structure thus resulted in Egen capturing only 44% of the original value of the program.

Two things became clear to Egen management. First, it was unlikely that it could sell the R&D program to Maxo at a value higher than what it was currently worth; second, it was likely that Egen would have to share some part of the $27 million value with Maxo for the deal to work. Egen felt that at least 75% of the value of the program should accrue to it and it would have to get at least $20 million for the contract to be fair (since the total value was $27 million). Egen went back to the model and made two changes (Figure 9.59). First, it set the milestone payments at the start of the expansion indications to $5 million instead of the $3 million suggested by Maxo. Second, it changed the royalty on all indications to 20% instead of the suggested 10% by Maxo. With these two changes, it revalued the contract for Maxo. It found that the value to Maxo with these changes was $7 million, allowing Egen to retain $20 million of the original value.

Maxo was a bit cool to the suggested changes by Egen. A week passed, and Egen management began to worry whether it turned Maxo off with aggressive negotiations. To its surprise, Maxo came back and suggested a new structure that looked very interesting. Maxo managers said that they were aware that

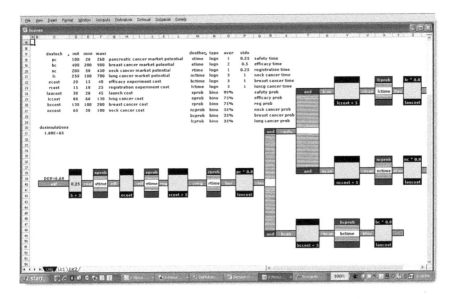

FIGURE 9.59
Evaluation of suggested new deal structure.

Egen was in need of cash in the short run. Maxo would like to increase the up-front fee to $5 million instead of the original $3 million, but wanted to keep the rest of the structure the same. Maxo also suggested that this was its last offer as they were in the hunt for similar candidates from other companies.

Egen management got together to evaluate the new offer. The $5 million up-front fee appeared very attractive as Egen was really in need of new resources to push its discovery programs forward. They could quickly calculate that the new proposal gave Maxo a value share of $13 million (the $15 million they calculated for the original contract minus the $2 million additional up-front fee offered). This was a 50:50 split of the value of the R&D program (with a total value of $27 million). It appeared to be a reasonable contract. Before agreeing, Egen wanted to take a look at the sensitivity of its assumptions on the value calculated.

One of the bigger indications for the candidate was breast cancer. Although it is a reasonably big market, the data from the preclinical studies indicated that the drug, if approved, would be administered in conjunction with existing therapies. Based on this, Egen evaluated the market and possible pricing. In calculating the expected value at launch for the breast cancer indication, it also considered that the new indication would only be approved 6 years from now (unlike the first indication, which was expected within 3 years). Egen wanted to know what impact the three assumptions on the potential of breast cancer (average, minimum, and maximum NPV) would have on the value of the drug. Figure 9.60 shows the impact analysis for the breast cancer assumptions.

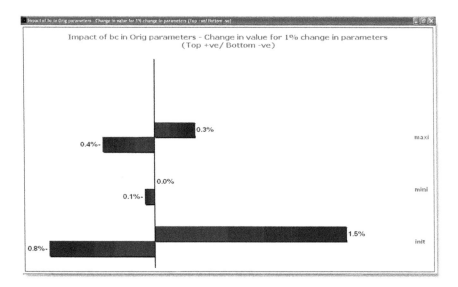

FIGURE 9.60
Impact analysis of the parameters of breast cancer indication on value.

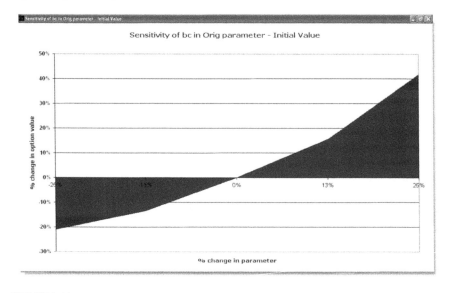

FIGURE 9.61
Sensitivity analysis of the expected value of breast cancer indication to value of the R&D program.

Egen also did a sensitivity analysis of the assumption of expected (initial) value of breast cancer indication to understand how it might affect the value of the R&D program. The result is given in Figure 9.61. The breast cancer assumptions affected the value of the R&D program close to 1:1 for errors

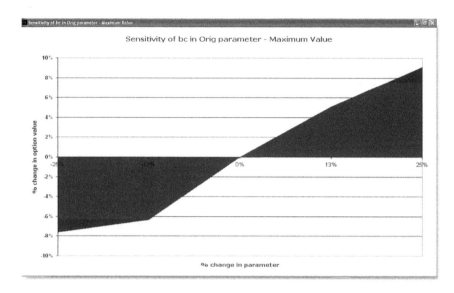

FIGURE 9.62
Sensitivity analysis of the maximum value of breast cancer indication to value of the R&D program.

within 15%. It was more sensitive on the upside and less sensitive on the downside. For example, a 10% error in breast cancer assumption caused a close to 10% error in the value of the R&D program. A 10% error in breast cancer value would represent an error of $40 million (10% of $400 million) in the estimates, and this would lead to a change of $2.7 million (10% of $27 million) in the R&D program value. Also note that Egen looked at errors here and not uncertainty. It already established that it did not precisely know the breast cancer indication's market potential. There is a range of possible outcomes from $200 million to $900 million. The estimation error of this range and its effect on the calculation of value were showcased by the sensitivity analysis.

Just to get a complete understanding, Egen also quickly performed a sensitivity analysis of the maximum value of breast cancer potential against the value of the R&D program and found that it was much less sensitive, as can be seen from the result in Figure 9.62.

Egen also wanted to run a few scenarios on the success rate assumptions for the efficacy experiment for the lead indication. It used both industry data and expert opinion to reach the 75% success rate assumption to see whether it could be a bit more optimistic here (say a success rate of 80%) and how that changed the valuation. Figure 9.63 indicates the results from the scenario analysis. Egen found that the value increased by 15% for a 7% increase in success rate (increase to 80% from 75%) or value increased 2:1 for a change in the success rate in the efficacy experiment. This reinforced what the company already knew: success rates obviously have a big impact on value. Success rates, however, cannot be changed much,

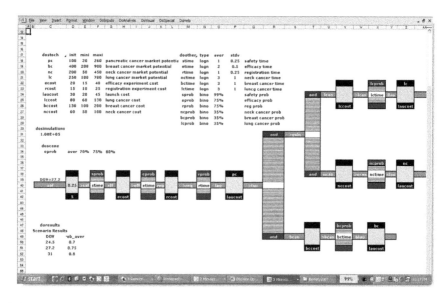

FIGURE 9.63
Scenario analysis.

especially if the failure emanates from toxicity. If failure is due to lower efficacy related to suboptimal pharmacokinetic properties of the chemical, perhaps it could be improved by a better dosage form. In this case, Egen felt that the 75% was a reasonable estimate of its chance to succeed in the efficacy experiment.

Overall, Egen management felt that the uncertainties captured were roughly correct. They also understood that decisions under these types of uncertainties were never perfect. The best they could do is to capture all the uncertainties using the best information they had and combine them systematically to reach a normative economic value. Use of such a process and value is the best and most objective way to reach decisions. After signing the contracts, the Egen CEO sat back, sighed, and remarked, "It is a tough business."

Selection of Optimal Legal Strategy

Consider a medical device company M-Device. It was founded 10 years ago by a couple of university professors and since then has grown to over 100 employees. Over the last decade, it has received multiple patents on some of the new technologies it invented. It has licensed some of these technologies to other companies for incorporation into devices manufactured by them. It also has a few R&D programs currently under development.

Last week was a busy one for M-Device senior management and the company's patent attorney. First, they got a letter from an inventor accusing the company of patent infringement. The inventor claimed that one of the devices currently in clinical trials by the company uses technology that infringes on a patent the inventor holds. The inventor wants compensation or threatens legal action against the company.

Later that week, the company learned that one of its competitors has a similar product in R&D and that the prototype may be incorporating technologies the company owns, infringing its own patent. If this technology is encumbered (if the claim of the individual inventor is true), then M-Device may not have a case against the competitor's use of their technology. But, on the other hand, if the inventor is wrong, M-Device may have a claim against the competitor. It is also possible that its technology supersedes the inventor's patent, in which case they may have a claim against both the inventor and the other company. M-Device's CEO asked, "How do we figure out the best strategy to pursue?"

M-Device decided to analyze its options as follows:

1. It could settle with the inventor, in which case it would not have any claim against its competitor using similar technologies. This strategy is named *settle*.
2. It could ignore the inventor's challenge (assume that it was irrelevant) and prosecute a claim against its competitor. It named this strategy pros1.
3. It can prosecute both parties (assuming its technology supersedes both). This strategy is called pros2.

Each of these strategies has an economic value associated with it. Obviously, the strategies depend on the value of the company's own development program and the private risk (probability) of winning in any of the actions taken.

M-Device first looked at the value of settling with the inventor and ignoring possible patent infringement by the competition (settle). Initial conversation with the inventor indicated that the inventor would be willing to settle for a structure that had an up-front payment and a settlement payment at a future time if and when the company received approval for its device and was ready to market it. M-Device estimated that the value of the product, if it succeeded in R&D and obtained FDA approval, would be in the range of $125 million to $300 million, with an expected value of $200 million. Based on historical data, it estimated that the probability of reaching the market (after considering both R&D and approval risk) was 75%. The cost of development that needs to be committed now was $50 million. If the device gets approved, launch costs would be $5 million.

If M-Device were to settle with the inventor now, it would have to pay the inventor $10 million immediately and then a settlement cost if the product got

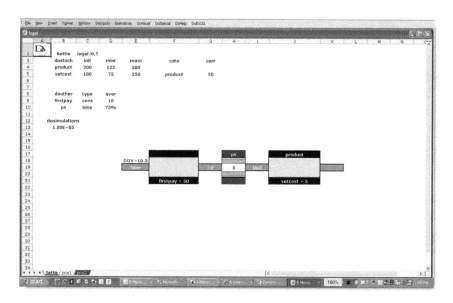

FIGURE 9.64
Model for settle.

to market. If the product did not reach the market (development and approval timelines add up to 3 years), there would not be any further settlement payment as the technology then would be rendered nonviable. M-Device estimated a future settlement cost of $100 million (range of $75 million to $150 million), roughly half the value of the product payable to the inventor if and when it marketed the product. It assumed that this settlement cost was correlated with the product value (a correlation of 50%) as the more valuable the product was, the higher the settlement was likely to be. The costs were not completely correlated as M-Device has some control over the negotiated settlement through its legal and IP strategy. Figure 9.64 shows the model and result. M-Device found that the settlement strategy could be valued at approximately $10 million.

With the same underlying assumptions on product costs, market potential, and success rate, M-Device analyzed the strategy of ignoring the inventor's challenge but prosecuting a patent infringement case against the competitor (pros1). It estimated that the legal fee would be approximately $35 million. It would only start the proceedings in 2 years as it gave the company sufficient time to assess the chances of the product's R&D success. It expected the case to drag on for a year after it started proceedings. It also gave itself a 50% chance of winning, and if it did, the settlement from the competitor would be around $75 million (range of $50 to $100 million). The company assumed that the amount of this settlement was also correlated with the value of the product (a 50% correlation).

However, ignoring the inventor's challenge had the downside of the inventor starting a patent infringement case against the company. M-Device

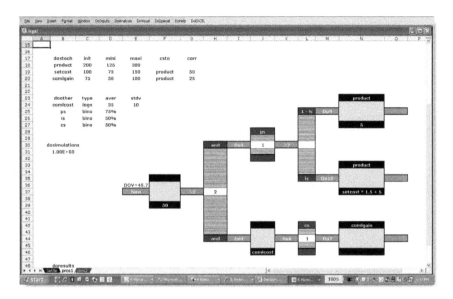

FIGURE 9.65
Model for pros1.

assumed that the inventor would do this only if the product reached the market. It gave the inventor a 50% chance of success on the legal case. If M-Device were to lose to the inventor, it estimated that the settlement cost would be 50% higher than what was estimated in the case of settling now. The model and result from this scenario are shown in Figure 9.65.

M-Device estimated that the value of this strategy was $46 million, nearly three times higher than the scenario in which it settled with the inventor now and ignored the possible patent infringement by the competitor. Buoyed by the increase in value of this strategy, it also considered a more aggressive strategy in which it prosecuted both the inventor and the competitor. It found that this strategy was also of similar value ($46 million). The model and result are shown in Figure 9.66.

To select the strategy it wanted to put in place, it analyzed the three strategies in different dimensions. First, it looked at the downside risk (probability-adjusted risk-neutral loss) against upside potential (probability-adjusted risk-neutral gain). The higher the upside potential, the better and the lower the downside risk, the better it would be. M-Device knew that from a purely economic perspective, it should pick the strategy that yielded the highest value (which is pros2). However, from a more practical perspective, it also wanted to minimize downside risk. Figure 9.67 shows the chart of downside risk versus upside potential for the three strategies discussed.

Although strategy pros2 (take legal action against both the inventor and the competitor) had the highest value, it also had the highest downside risk. The strategy of prosecuting only the competitor appeared optimal as it had

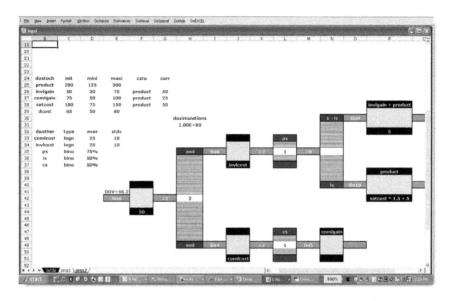

FIGURE 9.66
Model for pros2.

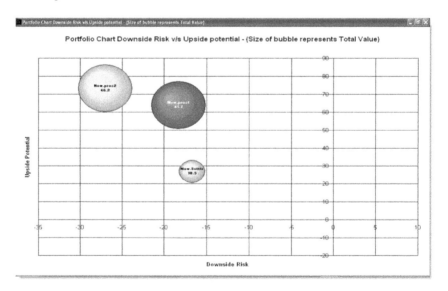

FIGURE 9.67
Value, downside risk, and upside potential of strategies.

similar value but much less downside risk. Although settling with the inventor now had the lowest downside risk, its value was significantly lower than the other two. Before deciding on a strategy to pursue, M-Device also wanted to see the value against total risk (standard deviation of risk-neutral value

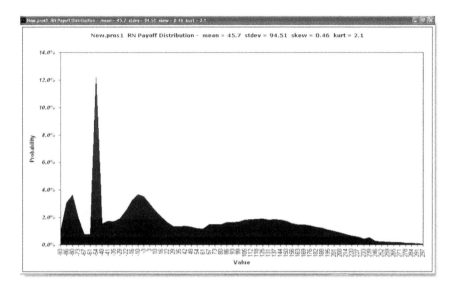

FIGURE 9.68
Risk-neutral payoff from strategy pros1.

distribution/average of risk-neutral value distribution). Figure 9.68 shows value against total risk. M-Device found that total risk of the two strategies (pros1 and pros2) was similar, so it decided to go with strategy pros1—take legal action against the competitor if the product succeeded and ignore the inventor's challenge for now (exposing itself to be sued later and incurring a much higher settlement cost).

Figure 9.68 shows the risk-neutral payoff for the selected strategy pros2. The peak represents scenarios in which the company committed the development cost but after 2 years decided not to proceed with any legal actions and the product failed. The negative outcomes are cases in which legal actions were taken, and either the product or the legal actions did not succeed. Because of the various events that may happen—technical failure and legal action failure—the outcome is quite variable, as illustrated by the risk-neutral payoff. Given all the information today (and all associated uncertainties), M-Device concluded that the strategy of initiating a legal action against the competition 2 years from now and ignoring the challenge by the inventor (not settle now, settle later if needed) is the most valuable strategy for the company to pursue.

Portfolio Management and Budgeting

Large pharmaceutical companies have hundreds of R&D programs in different stages. Most of them face significant uncertainty in market potential, costs, timelines, and success rates. Most need significant investments—people, money, space, and time—to progress through the R&D pipeline. Since resources are often limited, one of the fundamental problems in R&D is allocation of limited resources to many disparate investment opportunities to maximize the value of the company. This is a complex problem, and different companies approach it differently. Some use private risks as the primary proxy for selection and allocation decisions. Others may use minimization of cost, maximization of speed to market, or maximization of overall market potential as objective functions in selection and allocation decisions. None of these ensure that the portfolio that the company selects and advances is the most optimal from a shareholder value standpoint.

Multidimensional criteria-based selection and allocation decisions are also common in pharmaceutical companies. In this case, various criteria are given specific weights or importance. Each R&D program or investment opportunity is given a rank along each of the criteria dimensions. Such rankings may be created by a group of experts within R&D, such as the portfolio management group, or by a survey of the larger employee population. An overall score is then calculated for each R&D program as the weighted average of the product of criteria weight and rank. Decision makers may use such scores for project selection, resourcing, and prioritization decisions. Since multidimensional criteria-based scores are qualitative judgments—in both the importance of a criterion and the rank of a project on that dimension—they may not have any correlation to normative economic value. A significant amount of time and effort is wasted in the creation and use of such scores inside large companies in the belief that decisions are improved by such an exercise. Companies can both improve decision making and reduce the time taken in the process of project selection, resource allocation, and budgeting decisions by adopting metrics derived from economic value.

Figure 9.69 shows the result of a portfolio analysis of projects. The chart shows 16 projects with decision options value against traditional NPV. The size of the bubble represents option value. This chart shows the decision makers the following:

1. Which projects appear to represent the highest value for the company
2. Which projects have the highest option value (and hence higher uncertainty, decision flexibility, or both)

Such a depiction is useful to test and understand existing decision processes in the company. For example, if the company uses traditional NPV, either from DCF or from decision trees, a large number of projects in IP-intensive businesses (such as life sciences) and capital-intensive businesses (such as energy and high technology) will show negative values. If decisions are made strictly according to traditional NPV, only projects to the right of zero (on the x axis) can be selected. This means that only two projects in the portfolio have positive NPV. Immediately after such an analysis, results will be

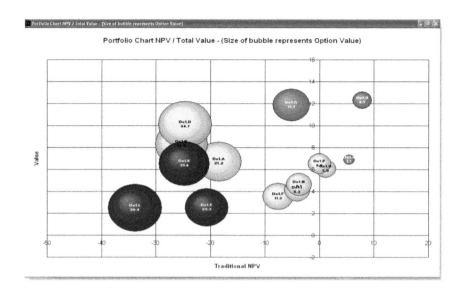

FIGURE 9.69
Economic value against traditional NPV.

discarded, and decisions will be based on gut feel. This is obviously a lot better as most projects in the portfolio have a positive value in this case. Projects higher up on the *y* axis (which shows economic value) are the best. If there is a resource constraint that does not allow the company to pursue all projects, decision makers can select a subset that has the highest value. When the company does not have the resources to pursue such projects, it may consider out-licensing them to other companies with lower cost structures that may find them attractive. Projects have a shelf life (primarily because of the limited patent life), so keeping them within the company (if resources are not available) may not be optimal.

In making a selection decision under a resource constraint, the company could also use downside risk as a cutoff measure. Figure 9.70 shows a decision aid that illustrates the downside risk and upside potential in each project. We define *downside risk* as the area under the curve of a risk-neutral payoff to the left of zero. Similarly, we define *upside potential* as the area under the curve of a risk-neutral payoff to the right of zero. Hence,

$$\text{Project value} = \text{Upside potential} - \text{Downside risk}$$

Decision makers may have risk aversion beyond a certain threshold downside risk in a project since a catastrophic failure of a project with high downside risk may result in the bankruptcy of the company. Although minimization of downside risk cannot be an objective function for project selection and resource allocation decisions, a certain threshold cutoff in downside risk always plays a role in such decisions. In the chart in Figure 9.70, projects

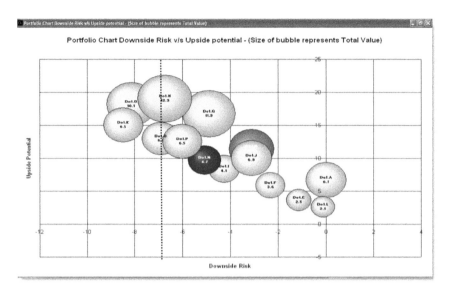

FIGURE 9.70
Value, downside risk, and upside potential.

are shown in a matrix of downside risk and upside potential. Decision makers may decide to elect a cutoff value for downside risk for each project. For example, in the chart, −$7 million is shown as the cutoff value for downside risk. In this case, projects K and D may be foregone despite their reasonably high economic value.

The company can also form an investment frontier in making a portfolio selection and budgeting design. Figure 9.71 provides an example in which the cumulative project value is shown against the cumulative start-up costs for the projects. The start-up costs represent the initial budget needed (this year) to initiate the project. In computing the economic value of the project, the model needs to incorporate uncertainty in future costs, timelines, private risks, and market potential in addition to the expected initial outlay. This is useful in selecting projects within a budget constraint. In the example in Figure 9.71, the budgetary constraint for the year is $255 million. By selecting the highest-value projects, the company can enhance shareholder value by approximately $80 million. In this case, it will forgo projects I, F, E, L, C, and M. The last two projects have negative economic value, and thus selection of these projects will result in a loss of shareholder value. Note that if the company were to select all projects (without consideration to economic value), its budget would have swelled to $450 million (from the current $255 million), and there is no net economic gain. The total economic gain remains approximately $80 million. Also, to execute the highest-value portfolio (of approximately $90 million) by selecting everything with a positive economic gain, the company needs a substantial increase in budget to approximately $400 million from the current $255 million.

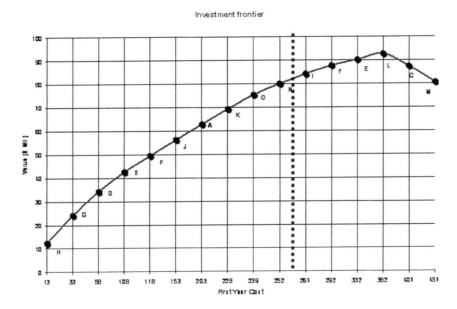

FIGURE 9.71
Investment frontier.

For the purely academic perspective, these decision aids may not be very compelling as the cleanest decision rule is to pick all projects with a positive NPV (economic gain). Unfortunately, in the practical world budgets exist, and managers are forced to select projects from a basket of available opportunities to meet a hard resource constraint. Using a holistic metric of economic value, downside risk, and upside potential, they can avoid the temptation of complex analyses of multidimensional criteria, weight, and ranking that do not correlate with economic value. Since such scores are qualitative and subjective by definition, they can easily be manipulated. Ultimately, resourcing decisions may favor those projects that have strong and vocal champions who can argue for a higher rank for their projects and demonstrate a higher score. The score becomes only a vehicle to show why a project is important. It can also lead to a long and contentious budgeting process as any score can be challenged. An objective criterion such as the economic value is less prone to such complications. Also note that traditional NPV is not useful in most cases as it does not capture the inherent uncertainty and decision flexibility in projects.

Economic value metrics, as demonstrated here, can also be tracked in real time as project characteristics change, and a portfolio adjustment is necessary. Annual budgeting and portfolio prioritization processes also do not allow dynamic changes in the portfolio as new information arrives. Portfolio management should be a dynamic process and not something performed at fixed time intervals such as a quarter or year. In many industries, the assumptions that went into the budgeting decisions at the beginning of the year may already have changed by the time the initial spending is authorized on the selected projects. In many cases, project plans are followed because "the budget was authorized." To make matters worse, incentives may be structured for managers to "come in on budget" rather than managing the projects better. Those who "managed" the budget perfectly may get a higher bonus. Rigid budgets and incentives based on accounting rather than economic value may destroy shareholder value rather than improve it. In long-cycle businesses such as life sciences, the effects of such value-diminishing decision processes surface only after a long period of time. By then, both the decision makers and the champions of qualitative budget processes move on and are not held accountable for the loss in shareholder value. A normative value-based dynamic capital allocation process is a necessary condition for success and survival of companies in industries with significant uncertainty and decision flexibility.

10

Case Studies in Technology and Manufacturing

Decisions to adopt technologies in companies—to improve productivity and profitability—are important from a shareholder value perspective. Often, such decisions have a critical impact on the viability and success of a company in a dynamic and hypercompetitive world. Technology has a big impact on the design of facilities and manufacturing plants as well as the management of logistics.

In the 1990s, many companies invested massively into information technology (IT)—in the collection, aggregation, storage, and reporting of data as in enterprise resource planning (ERP) systems and activity tracking systems (ATS). Often, such decisions were driven by fashion and fear. Many such investments were also made in fear of being left behind as technology improved economics of competing firms. Investments were aided by the promise of "huge" productivity improvements by software firms that created the technology and consulting firms that assisted in implementing them. The jury is still out on whether such investments reduced the flexibility of companies to adapt to changing conditions and diminished shareholder value. Software and consulting companies calculated a return on investment (ROI) for the implementation of their suggestions to convince their clients of the value. Such ROI calculations were made using deterministic calculations of costs, timelines, and benefits. As anyone involved in such implementations knows, costs, timelines, and benefits could not have been predicted with any level of certainty. In many cases, the risk of the technology implementation not delivering expectations was also ignored. Implementation projects were designed in a rigid fashion, allowing little decision flexibility after the project started. This is a valuable experience for all companies considering technology implementation without a clear understanding of how it affects shareholder value.

In considering technology and design improvements, it is also important to understand the impact on the entire system. In manufacturing, for example, an improvement in some aspects of the supply chain (such as reducing internal capacity to save costs) may have deleterious effects in other aspects (such as stock outs). Supply chain optimization techniques have been applied in many cases to improve performance, but the metrics used to measure improvement (e.g., cost, speed, availability, etc.) do not automatically imply an enhancement in economic value. Although reducing cost, increasing speed, and increasing availability (reducing stock outs) could be beneficial, it is impossible to say they are the correct actions to take individually without considering the holistic metric of economic value that incorporates all effects from the entire system. The systemic effects are not contained within the firm but extend to its customers and suppliers. How the company designs the interactions among its suppliers and customers has an important bearing in the performance of the supply chain. For example, contracts can be designed to increase flexibility, thus allowing the company to manage uncertainty better with a corresponding increase in economic value. Uncertainty is considered "bad" by traditional supply chain managers, and they focus on reducing uncertainty rather than increasing flexibility to manage it. As you have seen in the preceding chapter, uncertainty is not bad, but it is important to have the flexibility to manage it to enhance shareholder value.

In this chapter, we consider situations for which technology and flexible design could enhance economic value by allowing better management of uncertainty.

Manufacturing Capacity Optimization in Pharmaceutical Research and Development

Research and development (R&D) manufacturing in pharmaceutical companies faces significant uncertainty. Manufacturing is typically divided into two areas: drug substance and drug product. Drug substance manufacturing deals with the manufacturing of the active pharmaceutical ingredient (API). The API is then formulated into a drug product before it is used in human clinical trials. The API could also be used directly in animal experiments or in certain early human trials. Pharmaceutical companies can invest into creating API manufacturing capacity internally or outsource manufacturing to contract manufacturing organizations (CMOs). Creating internal capacity is costly as it often requires heavy up-front capital investment.

The demand for API manufacturing (we will use batches as a unit of measure) is a function of the R&D pipeline dynamic of the pharmaceutical company. It is a function of the stock of candidates in the pipeline, the position of candidates in the R&D process, and the flow of new candidates into the pipeline. If R&D is creating candidates at a healthy rate or the stock of candidates in the pipeline is high, the demand for API batches will be high. On the other hand, if R&D productivity is slow, demand for API batches will be low. R&D is also a creative process that may not work like a manufacturing process that produces at a constant rate. Ideas arrive randomly, and ideas progress up the R&D pipeline with uncertain timelines and success rates. This means that the demand for API batches is uncertain in terms of both quantity and timing. It is also the case that pharmaceutical companies experience "bunching of candidates" in the pipeline for a variety of reasons that result in a bolus of candidates at one time and corresponding increase in API demand followed by periods of low activity.

Investing in internal capacity takes committed investments up front. It will create a fixed capacity internally. Because it is internal to the company, the marginal cost of production of an API batch will be lower, after the up-front investment is taken. Creation of a fixed internal capacity, however, reduces flexibility for the company. If demand is low, the up-front investment expended in creating fixed capacity will be wasted.

A strategy of outsourcing API manufacturing will allow the company to match demand with capacity on an as-needed basis. Such flexibility is valuable in a highly uncertain environment. It, however, comes at a cost. The suppliers, CMOs, will demand premium pricing, and the marginal cost per batch of acquiring API from vendors will be higher than internal production. CMOs have made an investment in capacity on the basis of future demand and will demand a return on that investment. Since they can manufacture for many different customers, they have a higher level of flexibility in managing demand and capacity compared to a single pharmaceutical company. Thus, it is possible that the cost of acquiring an API batch from outside is lower than internal cost. Increasingly, low-cost countries such as India and China provide expertise and capacity for API manufacturing at a reduced cost, providing an additional lever of flexibility for pharmaceutical companies struggling to balance demand and capacity.

Let's study the case of GiantPharma. GiantPharma executives have been thinking about reconfiguring manufacturing capacity for the last several months. They have internal capacity for API manufacturing, enough to meet current demand. However, lately the discovery productivity has been increasing, and the pipeline is getting filled up again. They expect this trend to continue, and the current lull in the stock of candidates in the pipeline may change to a bolus. Looking forward, they anticipate a much higher demand for API manufacturing, more than what can be met with the current internal capacity. However, as has happened in the past, many of the drug candidates may not progress for technical reasons; this can also result in lower demand for API manufacturing. The number of batches needed is also a function of the chemical characteristics, the target disease, and the clinical experiment design. Thus it is impossible to predict the future demand with any level of certainty. The executives have two alternatives. First, they can build another manufacturing plant and thus increase internal capacity; second, they can pursue a strategy of outsourcing the excess demand to CMOs on an as-needed basis.

"This is complicated," said the chief financial officer (CFO). He has been wary of new investments in internal capacity, and as a veteran of the pharmaceutical industry, he has seen demand fluctuate wildly. Creation of internal capacity takes a huge investment up front, and that reduces flexibility. On the other hand, he also knows the perils of outsourcing when demand for API manufacturing is high industry wide. The lead time for new capacity is significant, and if the industry capacity is exhausted, prices will climb. Since

API batches are crucial for R&D, there will not be any alternative but to pay the price demanded by the CMOs.

One of the younger recruits in the R&D organization has been thinking about analyzing this problem. She decided to look at the alternatives from an options perspective. In discussions with experts in R&D, she pulled together the following information:

Current internal capacity	200 batches/year
Current demand	200 batches/year
Cost of new plant	$8 million
Capacity of new plant	100 batches/year
Variable cost of internal production	$50 thousand (K)/batch (average)
Fixed cost of internal production	$500K/year
Cost of on-demand outsourcing	$100K/batch (average)

To select a better strategy (internal capacity increase or on-demand outsourcing), she decided to model the problem considering all uncertainties. Since these are mutually exclusive strategies, she decided to assess the superiority of one strategy over the other. If internal capacity exists, the variable operating cost will be only $50K/batch. There is a fixed cost of $500K/year to operate the plant. The plant will incur the fixed cost regardless of demand. If the company builds internal capacity, it will save $50K/batch in variable cost of production per batch as the cost per outsourced batch is $100K. To do so, it has to invest $8 million up front to build the plant. The internal marginal cost of production does not include plant depreciation and any such accounting costs as once the investment is taken these are irrelevant for production decisions.

She decided to model the incremental benefit of internal capacity as an option to swap internal cost for outsourced cost. A *swap* is a recurring instrument (over a period of time), and at the elapse of a specific time duration, the owner will get the difference between asset and cost; this will be repeated until the end of the life of the swap. In effect, the benefit of internal production is the cost of the outsourced batch (the company saves that), and the cost is that of internal production (the company incurs that). The company currently has an internal capacity of 200 batches, so if demand is below that it will not outsource anything. After the new plant comes in line, the internal capacity will increase to 300 (the current 200 and the capacity of the new plant of 100), and if the demand is above that, the company will be forced to outsource the excess demand. After the new plant is in line, the company will meet demand up to 300 batches/year through internal production.

In studying demand patterns for the last several years, the analyst calculated the volatility parameter for demand to be 15%. Since the company has several decades of demand data, she felt that it was a reasonable estimate. She also decided to model the demand as a mean reverting function with a half-life of 5 years. Using industry-wide historical data, she found that periods of high demand are followed by periods of low demand and vice versa.

She speculated that the cyclical phenomenon is due to technology life cycles. Pharmaceutical R&D productivity is often driven by the invention of new technology platforms. Each technology platform allows scientists to explore hitherto unknown mechanisms and chemicals, leading to a bolus of drug prototypes. Over time, scientists across different companies and countries exhaust what can be discovered using the new technology, and the productivity drops until yet another new technology platform arrives on the scene. By studying the data, she determined that a half-life of 5 years was appropriate for the demand process.

In discussions with the company's experts in manufacturing, she gathered that both the internal and external costs of production are not fully predictable. The estimate of $50K/batch of variable internal cost is an average. Looking forward 10 years, some felt that they could see a range of $40K/batch to $65K/batch due to a variety of aspects, including manufacturing complexity, equipment and personnel availability, raw material costs, and raw material availability. Similarly, the cost of outsourced batches may also be volatile, in the range of $60 to $150K/batch with an average of $100K/batch as the CMOs are also subject to many of the cost uncertainties the company has and more. She also felt that the external cost/batch would be cyclical as contract manufacturing prices will be driven by the supply characteristics of the industry. If supply is limited and prices are high, more firms are likely to enter and push the prices down. Historical data indicated that the half-life of outsourced cost/ batch is in the range of 2 years. This is related to the lead time needed to build and validate plants that have to satisfy strict regulatory constraints in good manufacturing practices (GMPs). The internal cost of production is not driven by such cyclicality as it is not affected by external supply characteristics.

The analyst also thought about possible correlations. She felt that the internal cost of production was likely to increase as internal demand goes up. This is because the manufacturing has some shared components with other parts of R&D, and as space, equipment, and personnel become tight in supply (due to accelerating demand for all activities in R&D), manufacturing cost will likely go up. She decided to correlate the internal cost with demand at a 50% level. Based on expert opinions and historical data, she derived the following:

Internal demand volatility	15%
Internal demand half-life	5 years
Internal cost volatility	5%
External cost volatility	10%
External cost half-life	2 years
Internal cost correlation	50% to internal demand

The model is given in Figure 10.1. There is a fixed cost of $8 million up front (to build the plant) followed by a swap with a duration of 10 years with a swap frequency of 1 year. This means that the benefit to the company (the difference between external and internal cost of production) is calculated

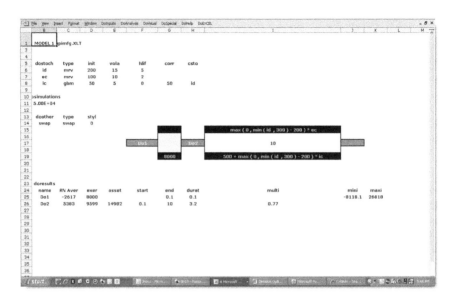

FIGURE 10.1
Addition of internal capacity in manufacturing.

every year based on the number of batches demanded. This process is continued for 10 years, the presumed lifetime of the plant.

The option to swap every year works in the following fashion. If the asset (in this case, the cost of outsourcing) is greater than the strike (in this case, the cost of internal production that includes both the fixed and variable components), we will swap asset for cost and take the difference (in this case, this is the saving from internal production). If the asset is less than the strike, we will not swap and take zero. This is similar to assuming that the company does not have any savings if the internal cost is higher than external cost, and it will simply outsource all production. This swap is not a plain vanilla swap, by which the difference between asset and cost is taken whether or not asset exceeds cost. This is a *swaption*, an option to swap. This means that the company has the option to decide whether to produce internally or outsource, depending on the cost characteristics.

Note that the objective here is to find out how much the company will save from the lower-cost internal production if it were to expand capacity. Although the initial estimate of internal variable cost of production ($50K/batch) is a lot lower than the external cost ($100K/batch), they are driven by different characteristics, and it is possible that in the future the cost of internal production will be higher than external cost. One such scenario will be excess external manufacturing capacity, forcing CMOs to drastically cut prices to keep their companies afloat. This may temporarily depress the external price. However, this will force many CMOs out of business, resulting in a reduction in external capacity and subsequent increase in external prices. Thus, the external cost per batch follows a mean reverting process just like the price of a commodity.

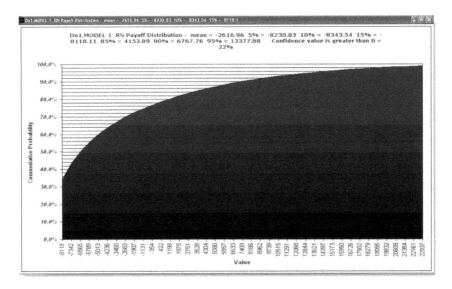

FIGURE 10.2
Cumulative probability of a risk-neutral payoff from a manufacturing capacity addition.

The model shows a $5.4 million savings from internal production over the 10-year period (using a real risk-free rate of 0). In Figure 10.1, the value of the swaption is designated as Do2. However, the capital cost of the new plant is $8 million, far in excess of the anticipated gain from lower internal production costs, making the net present value (NPV) of the internal capacity addition of −$2.6 million. Figure 10.2 shows the cumulative probability of risk-neutral payoff from the project. It shows that there is only a 22% chance of making an economic profit from adding internal capacity. It also shows that there is a 15% chance of losing $8 million, the entire capital investment.

This means that the company is better off employing a strategy of outsourcing on an as-needed basis, even though the cost per batch is significantly higher than the anticipated internal costs. In an environment in which demand, internal cost, and external cost are volatile, a flexible capacity (while costly on a per unit basis) is more dominant. The CFO was happy with the analysis. He asked manufacturing to devise a different plant design, one with lower cost and perhaps one in which the capacity can be added over time rather than in a single step. He felt that such a design could make the scenario of internal capacity increase more attractive.

The manufacturing manager felt that it was difficult to reduce the cost of construction of the plant according to the current design. She had been experimenting with a modular design that allowed capacity expansion incrementally. For example, the plant could be divided into four manufacturing lines (modules) that could be brought into production over time. This increased flexibility may enhance value and make the plant viable.

She was also thinking about the possibility of selling excess capacity to other pharmaceutical companies. If she is left with excess capacity, she can use it to manufacture API for other companies and create revenues, just like the CMOs. This way, she would not have to worry about overbuilding capacity. There are, however, some logistical and legal complications. One issue is the protection of intellectual property (IP) as the companies may be competing in the same area, and assurance of IP protection may make the process difficult to manage. So, a modular design may be the better approach for the company to consider.

By implementing a modular design, the up-front investment can be reduced significantly, and the company would have to take further investments in the future only if such investments added value. For example, if the demand increases rapidly and the external costs remain at the current level or even become higher, the company can add further capacity (at additional cost). In another chapter, the importance of such modular designs in manufacturing is analyzed.

Value of Modular Design and Flexible Capacity

Imagine a manufacturing plant with certain capacity to produce some end product, which can be a chemical (as in life sciences or commodity chemicals), an electronic component (as in technology), finished materials of a certain grade (as in mining), gasoline (as in a refinery), or any finished good in any industry. The capacity may be a function of equipment, people, space (for manufacturing and storage of raw materials, in-process inventory, and finished goods), and availability of manufacturing components (such as raw materials, additives, electricity, water, etc.). The capacity will require some lead time before it can be functional for the production of the finished goods. The term *raws* is used to represent any inputs into the process, *equipment* represents any conversion mechanism (including people), and *finished goods* represents the output from the process.

Managers of supply chains and manufacturing and demand/capacity balancing struggle to ensure that their capacity utilization is high enough for a profitable operation but not so tight it creates a delay in the system. Often, the lead times involved in creating capacity are long, so managers have to anticipate future demand well in advance and invest in creating optimum capacity for it. If they overdesign the supply chain, the result may be excess and idle capacity (increasing costs). If they underdesign, they may be forced to find alternative suppliers for excess demand (which cannot be met with existing capacity) or accept a delay in delivery (and related loss of revenue, increase in costs, or both).

In this case study, we first examine how much economic value can be added by closely matching demand to capacity. We assume that the equipment can be modular, allowing capacity to be increased in a certain fixed amount. We try to match demand to capacity by selecting the level of modularity (the capacity of equipment) that can be incorporated into the design. For the time being, the cost of modular design is neglected, and the sole focus is the benefit.

Let us study a manufacturing technology with a 5-year life. Before production starts, the manager has to make a decision regarding the design of equipment. The only variable considered here is the capacity of each machine. For now, assume that the cost of design is the same for each machine. The capacity per machine is selected up front. After that, each year the manager will decide how many machines to activate after observing the demand for that year. There are no costs to activate or mothball machines. Also, ignore capital costs completely just to understand how the ability to match demand to capacity affects value. The unit of production is the batch.

The model is shown in Figure 10.3. In this simple stylized example, we consider five stochastic processes: demand for batches (d), price per batch (p), internal capacity (i), internal cost per batch (c), and external (supplier) cost per batch (o). The company can produce a batch at a cost of c/batch if it has internal capacity of i batches. If it does not have internal capacity, it will incur a cost of o/batch by outsourcing the production to a supplier. All of these numbers are uncertain, but we have expectations of ranges for these variables over the timeline considered.

Each year, the manager has to make a decision to continue production. He or she must estimate the profits that can be created from operating the

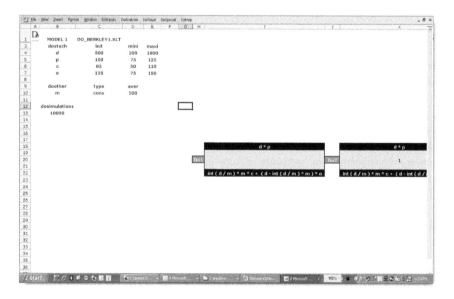

FIGURE 10.3
Modular manufacturing plant mode.

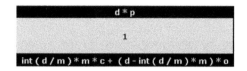

FIGURE 10.4
Production decision each year.

plant until its end of life. In any year, if the manager decides to discontinue production (because the sum of the current and future profitability of the plant is not positive), the entire facility is shut down, and all machines are scrapped. The price and costs are shown in thousands of dollars ($K).

The manager's decision each year is represented as an option as shown in Figure 10.4. Every year, the manager selects just enough machines to meet demand but not exceed it. The number of machines is represented by int(d/m), where int represents the integer part of (d/m), d is demand in batches, and m is the capacity of each machine. For example, if the demand is 200 and the capacity of each machine is 15, the manager will assemble the plant with int(200/15) = 13 machines. Since the manager does not want to overdesign the plant capacity, an internal capacity of 13*20 = 190 batches is accepted, and the remaining 10 batches are outsourced to an external supplier (at a cost/batch possibly higher than the internal cost).

The revenue is calculated as demand × price = (d*p). The cost is a function of internal capacity. The manager's decision to assemble int(d/m) machines means that the internal capacity for that year is set to int(d/m)*m. Hence, the cost of internal production is int(d/m)*m*c, where c is the internal cost of manufacture per batch. This, however, is not the total cost as the manager is forced to outsource (demand − internal capacity). This means that the cost of outsourcing is (d − int(d/m)*m)*o, where o is the outsourced cost/batch. Note that the decision the manager makes every year is related not only to that year's profitability but also to expected future profitability. If the manager chooses not to produce any year, the plant is shut down, and the manager effectively loses the option to produce in the future. Every year, the manager has to make a decision whether to continue the operation and if so how many machines to activate. There is the option to abandon the entire manufacturing operation in any single year.

Let us analyze the problem with a design decision of 500 batch capacity machines. Note that the demand expected per year is 500 batches (range of 200 to 1,000 during the course of the plant life), so selecting this "modularity" is a bit like a rigid plant (conventional plant built for a fixed capacity) as the manager will likely make a decision to utilize no, one, or possibly two machines per year.

Using the expected values and possible ranges of demand, price, internal cost, and external cost, we can simulate the asset and cost each year. The decision options problem has five sequential options; and we can solve this

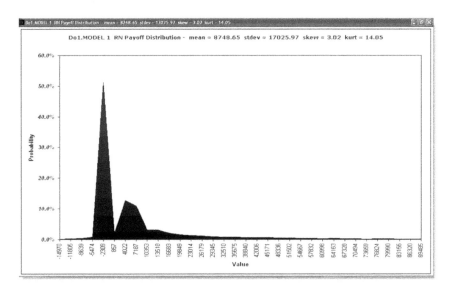

FIGURE 10.5
Risk-neutral payoff from a rigid plant.

entire problem using simulation and dynamic programming. Each year, demand, price, internal cost, and external cost are simulated, and optimal decisions are made based on what is known in that year. In doing so, the manager also has to project future years and combine those expectations with the current year's known profits (or losses). By analyzing the problem this way, the value of this rigid plant that has machines with large capacities of 500 batches/year can be calculated as approximately $9 million. Figure 10.5 shows the risk-neutral payoff from the plant. Negative values show plant production decisions at a loss (in anticipation of future profits) and subsequent abandonment of the entire plant because profitability did not improve as envisioned.

We can now run various scenarios in which the machine capacity differs. In each of the scenarios, we assume the design capacity of each machine to be fixed. The capacity of each machine has to be decided at design. Once this design parameter is selected, machines are produced only at that capacity. All machines are produced prior to the operating decisions every year. Operating decisions include whether the plant should be shut down and how many machines must be activated for production.

It is intuitively clear that the lower the capacity of a single machine, the higher the modularity of the plant and thus the higher the value of the plant as higher modularity allows closer matching of capacity to demand. Ten different scenarios starting from a capacity of 50 batches (microcapacity) to 500 batches (rigid capacity) are performed. The value of the plant as a function of the capacity of a single machine is shown in Figure 10.6.

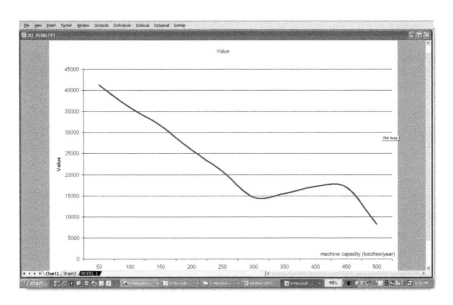

FIGURE 10.6
Value of the plant as a function of modular machine capacity.

The results show that the value of the plant continues to decrease until a machine capacity of 300 batches or so with a relatively stable period between 300 and 450 batches is reached. The reason that the value is slightly up from 300 to 450 batches is that the larger machines are able to match demand more closely. Remember that the demand is expected to be 500 batches, and the operating policy is to pick just enough machines so that internal capacity does not go over demand. When the capacity of a single machine is between 300 and 450, generally only a single machine can be picked. Thus, the larger the capacity, the closer the internal capacity is to demand with a single-machine plant. Since outsourcing costs are higher, the value of the plant is higher if the internal capacity is as close to demand as possible.

Now, let us make this problem more realistic. Assume that there is a cost of activation of each machine. One could imagine this as setup, shipping, installation, and maintenance costs that may increase with the number of machines and not with capacity. The cost of activation is assumed to be $600K per machine per year. In this case, the more machines used every year, the higher the cost of setup will be. Thus, there is an advantage for larger machines as there will be fewer machines and correspondingly lower activation costs. Smaller-capacity machines provide higher flexibility for the manager to match capacity to demand but also require higher setup costs. So, there is a trade-off between operating costs and setup costs.

Figure 10.7 shows the value of the plant as a function of capacity per machine from the analysis that includes the setup and maintenance cost per machine. The analysis shows that when the machine capacity gets too low

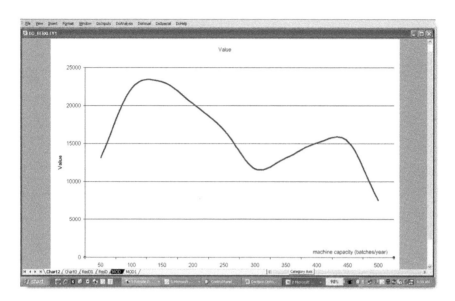

FIGURE 10.7
Value of the plant as a function of modular machine capacity when activation costs are present for each machine.

(fewer than 150 batches), the value from closely matching capacity to demand is counteracted by the setup and maintenance costs of machines. Since the costs of setup and maintenance increase with the number of machines, the smaller a single machine's capacity is, the larger the number of machines needed and the larger the overall setup and maintenance costs incurred. With such costs, the most efficient capacity is 150 batches per machine. We also get a secondary peak at 450 batches per machine as this allows closer matching of capacity to demand.

Figure 10.8 shows the risk-neutral payoff of a plant that considers the setup and maintenance costs of each machine and at a modular capacity of 150 batches/machine. The expected value of the plant is approximately $23.5 million.

In the previous analyses, we considered that if the manager makes a decision not to produce in any year, the plant has to be shut down, and no future production is possible. Now, consider a situation in which the manager can decide to shut down the plant any year but bring it back in a later year if the economics improve.

We first consider that the shutdown and start-up are costless, and the manager can select shutdown at the beginning of any year after having observed demand and assessed the value of operating that year. Since the manager can always bring the plant back in a future year costlessly, this is a less complex decision. The manager only has to evaluate whether positive profits are possible in any single year. If expected profits are positive in any single year, the manager will operate the plant, and if expected profits are negative, the

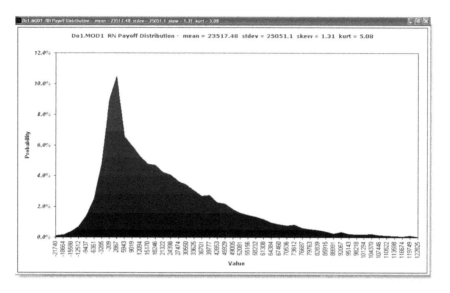

FIGURE 10.8
Risk-neutral payoff from the plant with optimal capacity modular machines with activation costs.

plant will be closed that year, waiting for the next year to see whether it is worthwhile to reactivate the plant. These decisions are now independent and can be made without consideration of future effects. We call these *switching options*. They give the owner of the options the opportunity to switch in and out (of production, in this case).

Figure 10.9 shows the risk-neutral payoff from the analysis (with a machine capacity of 500 batches/year) that has fixed activation costs per machine ($600K/machine) as well as flexibility to switch production on or off at the plant level costlessly. The value of the plant at a machine capacity of 500 batches/machine has nearly doubled to $17 million from the original $9 million. This is the additional value due to the plant-switching options. Also note that there are no negative profits as the manager will run the plant only if the profits are positive. In the other case, the manager had to make a judgment whether to operate the plant in any year at a loss in anticipation of higher profits in the future years. In some cases, such a decision would not have paid off, and the plant would have lost money in the future years as well.

Figure 10.10 indicates the risk-neutral payoff from the analysis (with a machine capacity of 150 batches/year) that has fixed activation costs per machine ($600K/machine) as well as the flexibility to switch production on or off at the plant level costlessly. High modularity coupled with switching options increased the value of the plant to nearly $26 million.

Figure 10.11 presents the value of the plant as a function of the capacity/machine when the plant-switching options exist. The chart shows that the

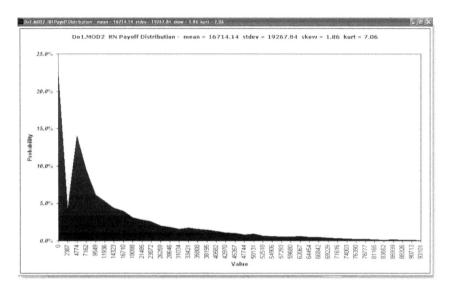

FIGURE 10.9
Risk-neutral payoff from rigid plant with switching options to shut down and restart.

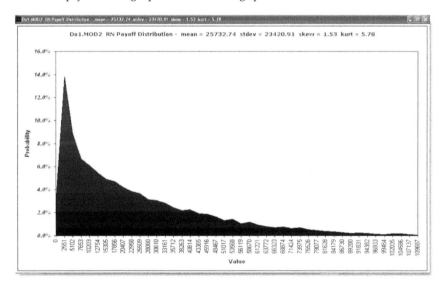

FIGURE 10.10
Risk-neutral payoff from a plant with optimal modular machines and switching options to shut down and restart.

ability to switch the plant on and off neutralizes some of the disadvantages of rigid machine capacity in high-capacity machines (as opposed to modular small machines). However, even with the plant-switching options, a modular capacity of 150 batches/machine appears optimal.

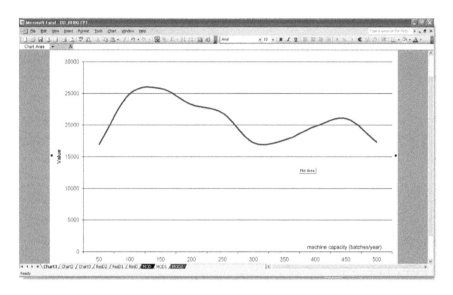

FIGURE 10.11
Value of the plant as a function of machine capacity when switching options exist to shut down and restart.

Now, let us consider the costs of shutting down, starting up, and maintaining the plant. As you can imagine, the shutting down of a plant will incur costs, including personnel costs. If the plant remains in the shutdown state, it will still require some maintenance costs. Restart also will require additional costs. Assume that the plant shutdown takes a cost of $3,000K, and the start-up requires $2,000K. While in the dormant state, the cost of maintaining the plant is $500K/year. The decision the manager makes each year now is more complicated. In some cases, the manager may operate the plant at a loss if he or she believes that the future profits will compensate for the current loss. In doing so, the manager also has to consider the alternative of shutting down, maintaining, and starting up the plant and associated costs. If the current year losses are modest, the manager may decide to continue and avoid the plant shutdown and start-up later as those costs may exceed the operating losses from a continuously operating plant. Figure 10.12 shows the risk-neutral payoff when start-up, shutdown, and maintenance costs of the plant are present at a machine capacity of 150 batches/year. Note that the value is approximately $2 million less than when plant start-up, shutdown, and maintenance costs did not exist.

In summary, it is important to consider flexibility in design, capacity, and operations of both the plant and machines to optimize the value of a manufacturing operation. Optimization based on conventional analyses (not considering uncertainty, not considering flexibility, or both) will not result in optimal decisions. Analyses such as the one discussed in this case study that considers uncertainty and flexibility in a normative economic framework

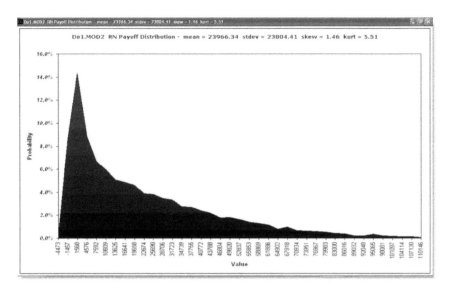

FIGURE 10.12
Risk-neutral payoff from the plant with switching options and costs to shut down/start up/ provide maintenance exist at optimal machine modularity of 150 batches/year.

provide optimal design and operating policies for manufacturing functions. Design decisions such as machine modularity and plant flexibility to shut-down/start-up; operating decisions to operate, mothball, or abandon; and configuration decisions such as number of machines in the operating plant can have significant impact on the overall value. These decisions cannot be made without a systematic consideration of uncertainty in all aspects: internal and external production, activation, shutdown, start-up, and maintenance costs as well as revenue.

Design of Outsourcing Contracts

Many companies are using outsourcing as an important flexible lever in managing and meeting demand. High internal capacity is costly when the demand is highly variable as it reduces flexibility in demand and capacity management. The discussions here are applicable to companies in any industry that have some internal capacity for the manufacturing of finished goods or intermediate components and supplement that capacity by contracting with outside suppliers to manage their uncertain demand. Increasingly, these outsourcing contracts have become an important strategic consideration for manufacturing functions as they have to consider many aspects such as risk of not meeting demand; internal and external costs (including marginal and total costs); cost of excess capacity; management; logistic complications of outsourcing; and the associated costs and economic gains. Often, such contracts are entered on an as-needed basis (*ad hoc* and tactical sourcing) or on a template for long-term contracts (with rigid contractual obligations). Because outsourcing is generally a procurement function, it tends to have characteristics of arm's-length transactions and thus eliminates opportunities for both supplier and buyer to introduce flexibility and thus enhance value through better management of uncertainty. Principles of procurement and negotiations currently practiced by large companies are remnants of the industrial revolution and are not very useful in today's world driven by uncertainty and the need for significant flexibility.

There are a number of features that both the supplier and buyer can consider in contracts, such as

Option to terminate

Option to delay

Option to accelerate

Option to increase or decrease committed quantity

Option to change delivery schedules

Option to prebuy certain quantity or capacity

Option to abandon sequestered capacity

Option to cancel or change orders

Consider a manufacturing company MF that has internal capacity to meet some of the expected demand for its finished goods. MF is considering entering into a contract with a supplier that will provide manufacturing services for batches that cannot be met with internal capacity. MF believes that this manufactured product will be active for 10 years until customers switch to something different. MS also has expectations for demand for this product. It is currently a stable 10,000 units, but looking forward 10 years, it sees a range of 7,500 to 15,000 units (represented as d). The price per unit today is $100/unit, but in the range of $90/batch to $125/batch (represented as p). MS has internal capacity of 8,000 units/year, but it could also change due to a variety of reasons, such as lower- or higher-than-expected equipment downtime, absenteeism, and a break in service. MS expects internal capacity to be in the range of 7,000 to 10,000 units/year (represented by i). Internal cost of production/unit (assume this is the marginal cost of production as the internal capital expense is already taken and thus is a sunk cost) is in the range of $70 to $90/unit with an expected cost of $75/unit (represented by c) and the expected profit (if manufactured internally) is $25/unit. MS also estimates the cost of an outsourced unit (if produced by an external supplier) is in the range of $85 to $120/unit with an expected cost of $95/unit (represented by o). Note that this is nearly 25% higher than the internal cost of $75/unit. All

FIGURE 10.13
Manufacturing model with simple excess demand flow through outsourcing and no fixed costs.

the ranges provided are minimum and maximum expectations in a 10-year horizon. Figure 10.13 is a representation of this manufacturing scheme.

We represent this problem as a 10-year plain swap. A *plain swap* is one in which the holder receives the difference between the "asset" (in this case, the revenue from selling the units) and "cost" (in this case, the cost of producing the units). A swap is a recurring instrument (over a period of time), and at the elapse of a specific time duration, the owner will get the difference between the asset and the cost; this will be repeated until the end of the life of the swap. In this case, the total duration of the swap (the manufacturing opera-tion) is 10 years. The swap duration is 1 year, meaning that the owner gets a payment at the end of every year. Note also that it is a plain swap. This means that the owner will get the difference between the asset and the cost whether it is positive or negative. During the operating life of the plant, in those years when the revenue is low (due to lower price) or cost is high (because of high internal costs, high external costs, high outsourcing or all of these) the prof-its can be negative. Profits will be positive otherwise. Figure 10.14 shows the risk-neutral payoff from such a manufacturing operation.

In this case, manufacturing decisions are made automatically regardless of new information. The company sets it up, makes an operating policy about internal capacity, and outsources everything the company cannot make. The manufacturer accepts the price in the market as well as its internal and external costs. In many cases, manufacturing companies are run this way. Under this scenario, the value of the manufacturing plant is approximately $2.1 million.

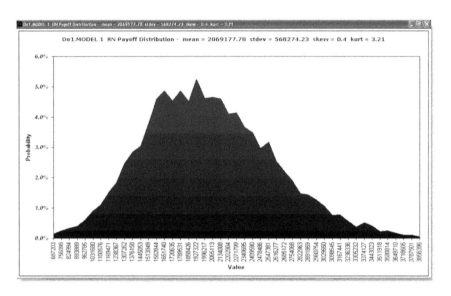

FIGURE 10.14
Risk-neutral payoff from the manufacturing plant with simple excess demand flow through outsourcing.

Let us introduce some fixed costs into this manufacturing facility to be more realistic. Assume that the fixed cost of running the facility is $600,000. Also, the marginal cost of production for the first 8,000 units is $0. The cost of producing the 8,000 units is already included in the fixed cost. For any production over and above 8,000 units/year, the marginal cost remains at $75/unit. A description of the model is given in Figure 10.15.

Figure 10.16 shows the risk-neutral payoff from the manufacturing operation with a fixed cost and initial sequestered capacity of 8,000 units/year. Note that the value of the operation is lower ($1.9 million instead of $2.1 million) than before as the fixed costs diminish the flexibility for the company in controlling costs.

The contract with the supplier currently is completely flexible. When an order is given, the supplier makes the supply and sends an invoice based on prevailing rates. The order is only given if the company has exhausted internal capacity. Now, assume that the supplier wants some assurances in terms of the size of the order per year. It wants an arrangement like a fixed-price contract. Every year, the company pays the supplier $190,000 as long as the order for units remains below 2,000 units/year. If the orders exceed 2,000 units, the supplier will charge $95/unit for the additional units produced. Figure 10.17 shows the model for this type of supplier contract. Note that the fixed cost of the company remains as before.

Figure 10.18 indicates the risk-neutral payoff from the analysis that has a fixed component in the supplier contract as well. Note that the value of the company dropped further ($1.5 million from the original $2.1 million)

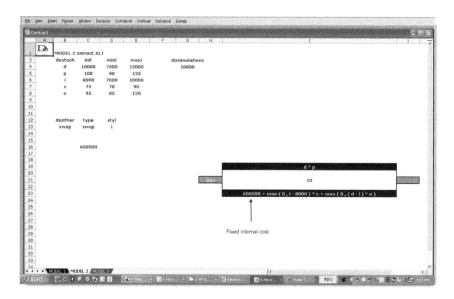

FIGURE 10.15
Manufacturing plant with fixed costs.

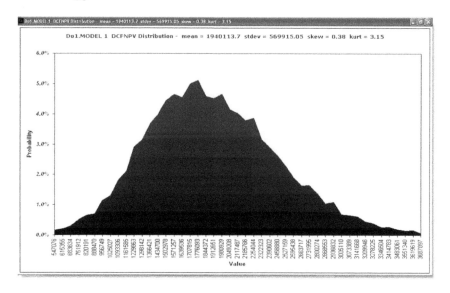

FIGURE 10.16
Risk-neutral payoff from manufacturing plant with fixed costs.

as flexibility to control costs diminished further. This contract is obviously beneficial to the supplier. One way to make this contract beneficial to the company is to negotiate a lower price for the sequestered capacity (fixed cost in the contract), lower the cost/unit charged by the supplier, or both.

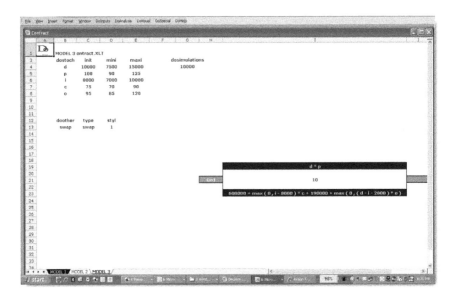

FIGURE 10.17
Plant model with fixed internal cost and sequestered external capacity.

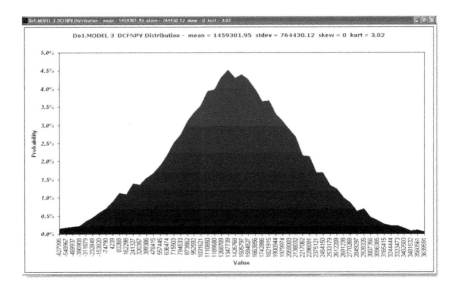

FIGURE 10.18
Risk-neutral payoff from the plant with fixed internal cost and sequestered external capacity.

Consider a contract in which the 2,000-unit sequestered capacity is bought for a cost of $190,000 but instead of the $95/unit for extra batches, the company negotiates a discounted price of $50/unit. The model and result are shown in Figure 10.19. This result will increase the value of the contract to $1.7 million.

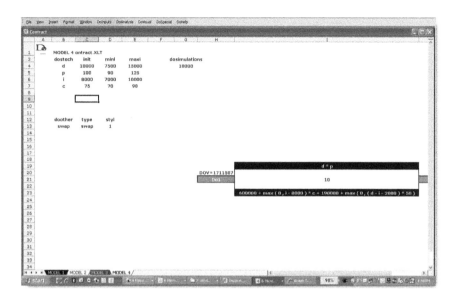

FIGURE 10.19
Manufacturing plant model with fixed internal cost, sequestered external capacity, and discounted pricing for excess demand flowthrough.

We could also consider a contract in which the company offers the supplier a higher cost/unit without the fixed fee. For example, suppose the company negotiates the cost/unit for the outsourced batch at a constant \$100/ unit (same as the selling price of the unit for the company) but avoids an up-front cost for sequestered capacity. It will pay \$100/unit for every unit it outsources. The model and result for such a contract are represented in Figure 10.20. The value of the manufacturing operation can be increased to \$1.85 million by avoiding the fixed cost even though the cost per unit is higher. In general, the more flexibility the company can retain in its supplier contract, the better.

We analyzed this manufacturing facility as a plain swap between revenue and cost. This means that there is no "optionality" in making the decision to manufacture. So, modeling demand, price, and capacity as stochastics have no relevance in the decision-making process. But, since the manager is not making any decisions based on observing them (as in the plain swap), it does not really matter. Modeling them as stochastic implies that they evolve in a smooth fashion, and that this year's demand is correlated with next year's demand. We could easily model them as probabilistic; in that case, this year's demand (or price or capacity) is uncorrelated with next year's demand (or price or capacity). We will be simply picking an independent sample from the distributions that define these factors every year.

Suppose the company has the flexibility to produce if the revenue is higher than cost and not produce otherwise; we can analyze it as an option

FIGURE 10.20
Manufacturing plant model with negotiated fixed cost for outsourcing.

to swap (swaption). The owner of a swaption has the right to exercise the swap at preset intervals. The exercise happens only if the asset exceeds cost. In all the cases discussed, if we treat the manufacturing operation as an option to swap, the value will be higher as it allows the company to shut down operations if profits are negative. Each option in such a swap is independent. The decision is made each year after observing the asset and cost. When decisions are made every year, they are not contingent; that is, this year's decision is independent of next year's decision, and the probability of exercising future options to swap has no effect on the current year's decision.

Postponement and the Value of Delaying Customization

NewTech Incorporated recently brought a new consumer electronics product to market, a wireless phone (WPhone) that can be used anywhere in the world and make phone calls over the Internet as well as on a regular phone line. Demand has been strong in test markets, and the executives at NewTech have been discussing better ways to organize the supply chain so that they can manage demand across the globe. For a new product like WPhone, it is often difficult to predict demand in various geographies. Because the base unit of WPhone is plugged into a power outlet as well as a phone line, it has to have different configurations for different countries. Countries have different power outlets, voltages, and phone jacks. This has caused a major headache for manufacturing and logistics staffs.

Marketing has been having problems projecting demand. Both the newness of the product and the various price points and geographies the product could be sold across have made it difficult to do this. The company could project an expected range of demand, but pinning it down to more precise estimates is nearly impossible this early in product introduction. Manufacturing's problem related to the fact that if they manufacture the product for one country (say, Japan), it will not be usable in another (say, India) as the countries differ in power and phone infrastructure. In essence, NewTech must manufacture a customized product for every country. If the product takes off in one geography and not in others, NewTech cannot switch shipments because of this customization.

A recent addition to the engineering department has been thinking about the problem and how to help the company. She felt that if the design of the WPhone is such that the ultimate assembly can be done closer to the consumer and later in the supply chain, the company can keep its options open. For example, if it ships generic units to a central warehouse in Asia and the customization components to the various country depots, it can manage uncertain demand better. If demand really takes off in India, NewTech can send the necessary base units to the local depot in that country and customize them for the Indian market. On the other hand, if demand is high in Japan, NewTech could divert the base units to that country, combine demand in Asia for the base units, and manage the production for the total demand.

The engineer decided to run a quick experiment. She studied a situation in which there were 20 country markets, each showing high variability in demand (say, lognormal distributions of an average of 10K units with a standard deviation of 5K, 50% of average). The demand from each country thus is highly variable. However, the combined demand from those countries may show a much lower variability. She ran a Monte Carlo simulation by combining the different demands and found that the coefficient of variation (standard deviation/average) in the combined demand was only 12%, a dramatic drop from the individual country variation of 50%.

She brought the idea to the manufacturing manager. The manager was not receptive first as he felt that the local customization would add cost to the end product. His experience told him that scale is important in manufacturing, and any time one does not take advantage of it centrally, it is likely to be more costly. But, the engineer was persistent. She argued that she understood that the cost would be higher, but the value of such a system would also be higher because of the higher flexibility. NewTech could direct products to areas where demand is high after observing the demand, and this would reduce stock outs. The manager reluctantly agreed that she could compile the necessary data to conduct an analysis.

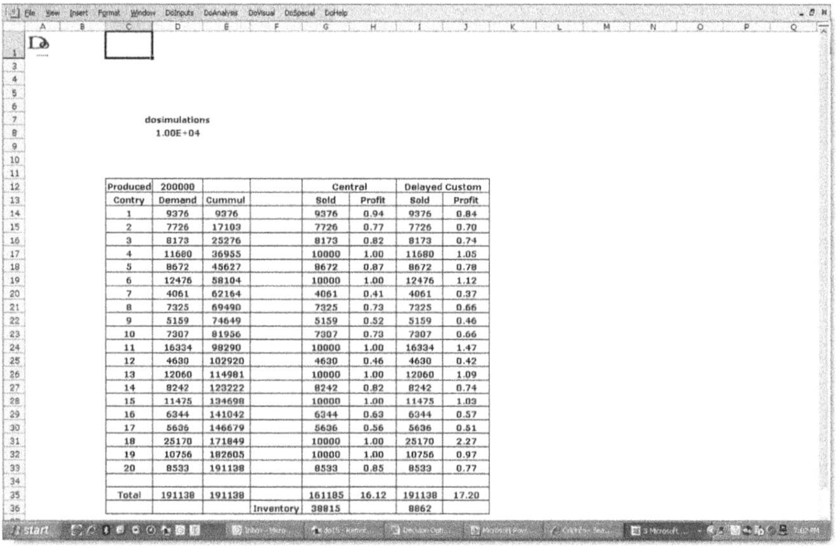

dosimulations
1.00E-04

Produced	200000			Central		Delayed Custom	
Contry	Demand	Cummul		Sold	Profit	Sold	Profit
1	9376	9376		9376	0.94	9376	0.84
2	7726	17103		7726	0.77	7726	0.70
3	8173	25276		8173	0.82	8173	0.74
4	11680	36955		10000	1.00	11680	1.05
5	8672	45627		8672	0.87	8672	0.78
6	12476	58104		10000	1.00	12476	1.12
7	4061	62164		4061	0.41	4061	0.37
8	7325	69490		7325	0.73	7325	0.66
9	5159	74649		5159	0.52	5159	0.46
10	7307	81956		7307	0.73	7307	0.66
11	16334	98290		10000	1.00	16334	1.47
12	4630	102920		4630	0.46	4630	0.42
13	12060	114981		10000	1.00	12060	1.09
14	8242	123222		8242	0.82	8242	0.74
15	11475	134698		10000	1.00	11475	1.03
16	6344	141042		6344	0.63	6344	0.57
17	5636	146679		5636	0.56	5636	0.51
18	25170	171849		10000	1.00	25170	2.27
19	10756	182605		10000	1.00	10756	0.97
20	8533	191138		8533	0.85	8533	0.77
Total	191138	191138		161185	16.12	191138	17.20
			Inventory	38815		8862	

FIGURE 10.21
Monte Carlo simulation of centralized and delayed customized manufacturing.

The engineer collected the following data:

Cost per unit with central manufacturing	$100
Cost per unit with decentralized customization	$110 (10% higher)
Revenue per unit	$200

The company has excellent currency-hedging programs in place, so for this simple analysis, she did not worry about currency effects. She put together a simple model as shown in Figure 10.21. The model's table shows the following:

Produced	200K units
Country	20 different countries
Demand	Simulated demand in each country: lognormal(10k, 5K)
Cummul	Cumulative demand to a maximum of 200K units (total produced)
Central	Units sold and profits from each country with an allocation of 10K units to each country; profit per unit = $100 ($200 revenue − $100 cost)
Delayed custom	Units sold and profits from each country with no specific allocation of units to any country; profit per unit = $90 ($200 revenue − $110 cost); the cost includes 10% increase due to decentralized customization

Figure 10.22 depicts the distribution of profits from a simulation of centralized manufacturing. The expected profit is $16 million.

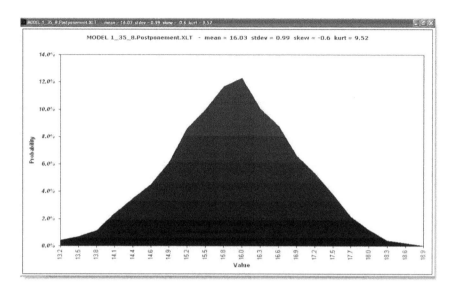

FIGURE 10.22
Risk-neutral payoff from centralized manufacturing.

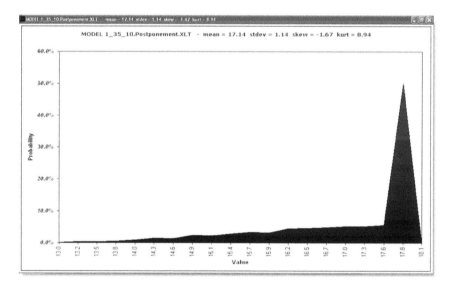

FIGURE 10.23
Risk-neutral payoff from delayed customization.

Figure 10.23 shows the distribution of profits when delayed customization is implemented at a 10% additional cost. The profit of $17.1 million is 7% higher than in the case of centralized manufacturing (without delayed customization). The profits are higher (in spite of the higher costs) because of the company's ability to more closely match demand wherever demand exists.

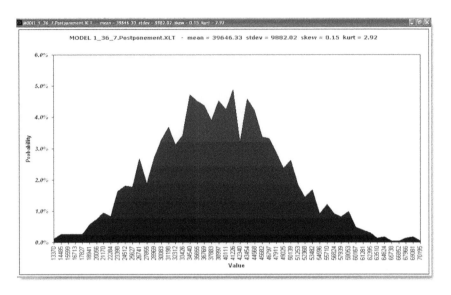

FIGURE 10.24
Unsold inventory in centralized manufacturing.

The increased profit from delayed customization can be understood from the inventory that remains in both systems. Given in Figure 10.24 is the inventory without delayed customization. Nearly 40K units (20%) of the produced units could not be sold because of variability in demand in different countries and the policy of early customization and fixed allocation to each country. Figure 10.24 presents the unsold inventory with centralized supply chain.

With delayed customization, inventory can be reduced significantly, as shown in Figure 10.25. The inventory is only 10K units, about 25% of the case without delayed customization. Conventional wisdom in scale-based manufacturing and logistics needs to be rethought when uncertainty is present. Introducing flexibility in the supply chain may be strategically important for many companies. Such flexibility can come from a variety of sources, including the design of products, plants, logistical systems, and contracts.

The supply chain is a significant strategic issue for manufacturing and logistics companies. Many companies attempt to reduce uncertainty for improved performance of the supply chain. Although information that allows better prediction (and thus lower uncertainty) is good in the management of complex supply chains, the economic value gained from introducing flexibility should not be discounted. Companies such as Wal-Mart and Hewlett-Packard Company have long recognized this and have invested in increasing flexibility. In many cases, the success and failure of companies may depend critically on how they design manufacturing and logistical systems as well as the features they incorporate into supplier and buyer contracts. Also, it is not only the functionality of products that is critical but

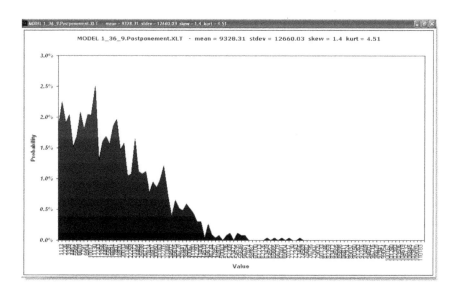

FIGURE 10.25
Unsold inventory in delayed customization.

also their industrial design. Ability to delay decisions and scale up through modularity are critical attributes of industrial design.

Flexible Manufacturing Assembly Line for Automobiles

AutoCar, a large manufacturer of passenger cars, has been having a difficult time lately. Gas prices have been volatile, and consumer preferences for automobiles are more unpredictable every day. The design department has been busy putting the final touches on two new models: a sports model called Zephyr and a hybrid intermediate called Electra. Both are revolutionary designs and will require building new assembly plants as both the body and the components are of new materials, and these models are manufactured very differently from existing models. The design department has been careful to design them to share certain components and materials as they have been very aware of the significant unpredictability in the market both for demand and preferences for types of cars.

For the new manufacturing plant, AutoCar is considering two different designs. Both designs have two assembly lines. In design A (called fixed), each assembly line is specialized. That is, each line can only make one type of car. Line 1 will make Zephyr, and line 2 will make Electra. In design B (called flexible), assembly lines are not specialized, and both line 1 and 2 can make either car model. The flexible design is more expensive and will require a higher level of investment than the fixed design. All previous manufacturing assembly lines at AutoCar have been versions of the fixed design. All lines are specialized and can make only one type of model. The biggest advantage of the fixed design is cost. It is cheaper and easier to design.

Given the cost advantage, some members of the executive committee questioned the consideration of the flexible plant design and consider it just a waste of time, money, and space. They feel that although the future is unpredictable, the flexibility offered by the other design is unlikely to justify the cost. The debate has accelerated as a small cohort of young plant managers started pushing for flexible designs. They feel that it does not make sense to have specialized lines for single models as the demand for them is impossible to predict. The utilization of specialized lines will be less, they argued, and that alone may justify the additional investment.

As there was no obvious way to select from the alternatives, the CFO decided to devise a model to consider the uncertainties and production flexibilities. Both assembly lines each cost $350 million to build for a total investment of $700 million. If they are designed such that either line could be used for the production of either type of car, the cost per line goes up by $35 million (10% more), and the total required investment will be $770 million.

The marketing department felt (based on its research) that they could price the traditional gas-powered vehicle at around $20K. The range of pricing will be $18K to $23K, and the actual price point will depend a lot on competition, consumer attitudes, and fuel prices. The prices also will have to change during the course of the vehicle's lifetime. Analyzing the historical data of similar vehicles, marketing felt that the range was reasonable. For the hybrid vehicle, they were confident that a premium could be charged, and they estimated this premium to be $5K per vehicle. Again, the ultimate pricing will depend a lot on consumer attitudes and fuel prices and other alternatives available. At these price points, marketing felt that the demand for each vehicle would be in the range of 100K units/year. Demand is notoriously unpredictable, but they had historical data (on both their own vehicles and competitive vehicles) to reach a rough estimate of demand at 75K to 150K units/year for each (gas powered and hybrid). Younger analysts felt that selling over 100K units was almost a certainty, but they eventually yielded to the veterans, pointing out that they had been equally optimistic last time around, and the result was not the way they envisioned.

The design department created blueprints of the assembly line with a capacity of 100K vehicles/year. In consultation with the plant manager, the CFO estimated that the utilization would be about 90% (accounting for plant shutdowns, strikes, and other unforeseen events). It appears reasonable to assume a capacity of 85K to 100K vehicles/year from each assembly line. The CFO then met with the accounting and design departments to get an estimate of the cost of production. The designers had used some new materials (carbon fiber), and they had to estimate the production capacity of the suppliers of those materials and the price points they could negotiate with them.

The analysis showed a cost of the vehicle (excluding any capital costs) at $19K for the gas-powered one and a lofty $23K for the hybrid, driven by batteries and other design components. Both showed net margins of 10% (excluding capital costs).

Based on information from the design and marketing departments, the CFO also summarized the following inputs:

Price of the traditional gas-powered vehicle	$18K to $23K
Price of the hybrid vehicle (with batteries)	$23K to $28K
Capacity of each assembly line	85K to 100K vehicles/year
Demand for each type of vehicle	75K to 150K vehicles/year
Cost of production for gas-powered vehicle	$17K to $21K
Cost of production for the hybrid	$21K to $26K
Life of the plant	8–12 years

The life of the plant was set at 10 years, but some flexibility exists to extend it to as long as 12 years. There may also be unforeseen circumstances that may force the company to close the plant prior to its expected life due to technology obsolescence, regulation, environmental issues, lack of raw materials at the specific location, currency appreciation forcing location change, and a host of other factors.

The CFO also spent time with the planning-and-scheduling department to better understand how they will operate the plant. In both designs, they will have the same operating policy. At the beginning of each month, they will assess the price the market will bear, compare that with their costs, and make a production decision. If the price is lower than cost, they will shut down the plant, and if it is higher, they will operate. Using just-in-time techniques, they have reduced inventory almost to zero. The shutting down and starting up are costless as the plant is almost fully automated with robotic technology. There will only be three employees in the entire plant (one each for the two assembly lines and one safety inspector), and the cost of employees is negligible in the cost of production.

To determine whether it is worthwhile to make the $200 million additional investment to design and build a flexible plant (that allows production of either type of vehicle), the CFO decided to value both designs. First, the fixed plant was modeled as shown in Figure 10.26. The inputs in the model include price (denoted as Pg and Pe for the gas-powered vehicle and the hybrid. respectively); manufacturing capacity (Mg and Me, respectively); demand (Dg and De, respectively); and cost (Cg and Ce, respectively). None of these is precisely known, but estimates are available. The CFO modeled all inputs as geometric Brownian motion (GBM) and assumed a real risk-free rate of zero.

Each assembly line is modeled as a swaption in which the swap interval is 1 month. However, the life of the plant is also an unknown. The CFO modeled this as a probability distribution with an expectation of 10 years (but in

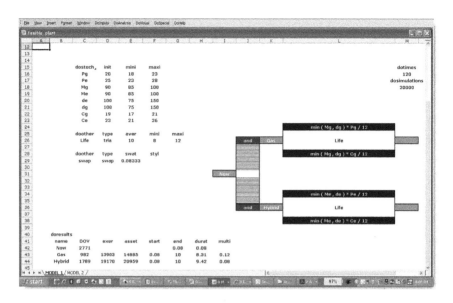

FIGURE 10.26
Automobile assembly plant with rigid design.

the range of 8 to 12 years). A swaption replicates the ability to produce when the price is higher than cost in each month. The total value for the company is the summation of value from both assembly lines. In each assembly line, the production is set as the maximum of capacity or demand. If the demand is lower than the attainable capacity of the assembly line in any month, the company will produce only the demand and leave the rest of the plant idle. The company has no cost for idle capacity. If the demand is higher than attainable capacity, the company will produce as much as it can of that type of vehicle in its assigned assembly line. If the demand is high for one type and low for another, only one assembly line will be run to capacity, and the other line will have idle capacity. The CFO estimated that the value of the fixed plant was $2.77 billion.

Next, the CFO modeled the flexible plant. The operating policy remained the same as for the fixed plant. The difference was that the operating managers can build either type of car on either assembly line. Every month, they will assess the total demand (for gas powered and hybrid). First, they fully allocate production of the hybrid demand as the margins are slightly better on it. Then, they assign the remaining capacity on either assembly line to the gas-powered vehicle. They will produce the total demand for both vehicles if the attainable total capacity for both assembly lines is higher than total demand. If the total demand is higher than total attainable capacity, they will likely meet the demand for the hybrid but not necessarily for the gas-powered vehicle. The model for the flexible design is shown in Figure 10.27. The equations on the swap constructs show the production policy of allocating

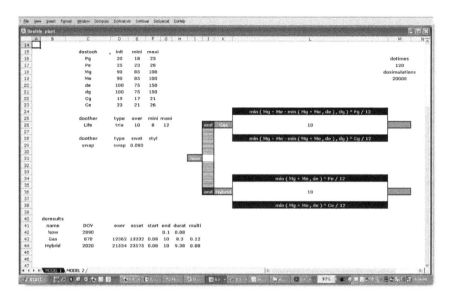

FIGURE 10.27
Automobile assembly plant with flexible design.

capacity to fully meet hybrid demand and diverting the rest of the capacity to gas-powered vehicles. Also shown are decisions made every month after observing demand, capacity, cost, and price.

The analysis shows that the flexible design is worth $2.89 billion, $119 million more than the fixed design. The value of the flexibility, allowing production of either vehicle on both assembly lines, thus is $119 million.

Given that the incremental cost of adding this flexibility is $70 million, the CFO found that it is worthwhile to add flexibility. In doing so, the company could enhance a net $49 million in shareholder value. Armed with the analysis, the CFO was able to build consensus with the executive team that the flexible design was a better one to pursue.

Information Technology Implementation in Stages

GoTech executives have been debating over the implementation of a new IT system to improve the productivity of employees for several months. It all started when some of them attended an IT conference where a large software company exhibited its Productivity Enhancement System (PES) to improve workplace productivity. The company showcased ongoing implementation at several of GoTech's competitors. This worried the chief information officer (CIO) of GoTech, who felt that if the competitors implemented the system, it may be important for

Gotech to take the plunge as well. The CFO was against it and remained convinced that the massive up-front investment to implement the technology was a huge risk as they did not have much data to show that PES in fact improved productivity.

At the conference, a PES executive described how important it is for companies to implement the system across the entire enterprise. The executive showed that an enterprise-wide implementation lowered licensing fees (on a per seat basis) as the company would be purchasing a large number of seats together. PES pricing was a function of how many copies the client purchased in one implementation. PES had scale advantages in terms of people and equipment, and the pricing included implementation. The executive showed statistics that indicated that many of the PES clients decided to implement the system across their entire companies to take advantage of the favorable pricing.

"I am not really sure," said the CFO. "How do we know if PES is effective and how much productivity improvement can we expect? We have 100,000 employees and five different departments across a dozen countries, and it will be a big investment for us up front. If the expectations are not realized, it can have a significant impact on our stock price." The CFO also felt that the fact that the competition was implementing PES was not sufficient for GoTech to take the plunge.

In a recent management meeting with the CIO, one of her technology managers suggested that perhaps it may be better for GoTech to try a staged approach to this implementation. The manager suggested a sequence of implementation projects, each project targeting a portion of the employees.

The technology manager decided to run a quick analysis. One of the issues the manager wanted to investigate was the benefit from the implementation. PES was a new system, and although some companies were currently going through implementation, little hard information was available on how much benefit they would derive from it. An associate suggested that data from the implementation of similar enterprise systems may be useful in this regard. The associate identified a mature technology implemented by hundreds of companies in the last decade called TRS as a close proxy of PES. TRS (Time Reduction System) reduces the time spent by employees in certain tasks and thus improves the productivity of an enterprise.

TRS has been around for nearly a decade now. There are many companies that have adopted it. Because companies implemented the technology at different times, some adopting it early and some late, the associate has been able to identify a cross section of companies with differing duration of use of the technology. He grouped the companies into buckets, differentiated by their use duration. He found companies to cover the entire spectrum, some starting off and some having as much as 5 years of use. He also interviewed representative companies in each bucket to understand their experience with the technology and found that the benefits companies gained from the implementation of TRS varied widely. The more experience they had with the system, the more varied their experiences had been. Some companies were extremely happy with it, and they indicated that the technology met their wildest expectations. On the other hand, some companies felt that they lost money from the implementation as they failed to capture any productivity gains. In talking with companies that had at least 5 years of experience with TRS, the associate found that the happiest company with TRS was able to derive benefits three times its cost. The most disappointed company suggested that the benefits were almost zero, meaning that it could not even recoup costs. There were many companies with benefits between these extreme cases. Based on this, the GoTech associate calculated a volatility of expected benefit of 30% for PES.

On the cost side, he found cost overruns of as much as 50% driven by unexpected complications with software, training, and databases. Some companies completed the project at only 75% of the initial expectation. These data allowed him to calculate a volatility factor of 10% for cost. Each implementation lasted between 9 and 18 months (average of 12 months), and timelines were unrelated to the total costs.

The GoTech associate wanted to derive an expected ROI for various project plans. He considered five different project plans. In the first plan, the implementation would be in a single stage for the entire company. In plan 2, the company was divided into two groups, and the implementation was divided in two sequential stages. In plan 3, the company was divided into three stages with sequential implementation, and so on. He also assumed that the benefits, gains, and timelines from the sequential stages in each of the plans were correlated. For example, if cost overruns happened in one stage, it was likely they would happen again in subsequent stages. If benefits derived from one stage were high, they were likely to be higher in subsequent stages as well.

The results from the analysis are shown in Figure 10.28. The figure shows that the higher the number of stages, the higher the ROI. This is because a staged implementation provides higher flexibility. As the company goes

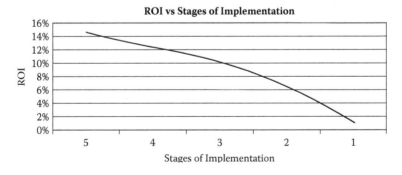

FIGURE 10.28
ROI versus stages of implementation.

through the stages, it learns more about the benefits of the technology and cost of implementation. This provides the company with an option to abandon future implementations if the benefits do not exceed costs. For example, if cost overruns are high and the derived benefits from the first-stage implementation are low, the company may forgo further implementations. On the other hand, if the costs are low and benefits are high, the company will continue implementation in subsequent phases. The more stages there are, the higher the optionality and the higher the expected ROI from the project at start.

11

Case Studies in Commodities

As the world copes with rising costs of conventional energy-producing commodities such as oil and gas, alternative energy technologies are becoming increasingly viable. These include solar, wind, geothermal, and biofuels. As transportation and efficiency considerations also become critical, traditional delivery mechanisms for energy are also being rethought. Electricity is gaining momentum as a method of delivery as electric automobiles and trains gain popularity. Storage of energy is also becoming critical as energy users move more into alternative fuels. The industry thus is experiencing unprecedented and massive changes in all aspects of energy production, transmission, delivery, and storage.

Commodities that produce energy such as oil, gas, and agricultural inputs (as in biofuels) share many common characteristics with other commodities such as metals, construction materials, water, and synthetic chemicals. In all cases, the price of the commodity is driven by supply and demand. Supply and demand are influenced by economic cycles and, in certain cases, the perceived stock of raw materials in the world. With the economics improving in the BRIC (Brazil, Russia, India, and China) and other developing countries, unprecedented global integration in production, pricing, and consumption of commodities and the availability of a plethora of financial instruments for hedging and speculating have resulted in complex price dynamics and increased volatility. Decisions involving the discovery, production, or manufacturing and logistics of commodities are increasingly complex, requiring considerations of choices among substitutes of fuel, commodity inputs, and consumption modalities as well as flexibility in timing and quantity.

Risk management is also of great importance for these markets. Enterprise risk management has to consider real and financial risks in a single framework as they are closely integrated. Portfolio management of real and financial assets requires considerations of interactions among them as well as the capital structure and operating flexibility of the company. Thus, decisions made by different entities in an organization such as corporate finance, contracts, procurement, and engineering cannot be made independently any longer as the viability of the company depends on the holistic management of all components of risk and economic value.

In summary, in all segments of the energy and commodity markets, decision complexity has increased, aided by interactions and convergence among components, globalization, volatility, and the need for holistic risk and portfolio management. Analytical tools that aid in the following aspects of business are now requirements for better management of companies engaged in energy and commodities:

1. Modeling of decision problems taking into account volatility and other characteristics in price and cost processes, choices among converging inputs and outputs, as well as flexibility in timing/quantity, leading toward decisions based on economic value

2. Assessment, evaluation, and management of various types of risks and incorporation of such risks in decision frameworks in establishing normative economic impacts of decisions

3. Forecasting of parameters that drive fundamental processes (such as price and cost) from historical observations and proxy analysis

4. Holistic portfolio management, taking into account interactions among real and financial assets as well as the capital structure of the firm

Decision makers have to select and design value-maximizing projects and manage an optimal portfolio. They also have to conduct enterprise risk management, optimization of supply chains, financial engineering, and technology design, all in the context of a normative economic value. This will result in better and faster decisions, communications, and consensus building in all aspects of production, logistics, and consumption.

As discussed, commodities such as metal and crude oil have supply-and-demand characteristics that force the market prices to exhibit mean reversion. *Mean reversion* is the tendency for prices to move toward long-run equilibrium prices. When prices move up from the equilibrium level, supply increases (because of higher investments by suppliers), and demand decreases (because of lower consumption and higher conservation by users), effecting downward pressure on price. Similarly, when prices move down from equilibrium level, supply decreases (because of lower investments by suppliers), and demand increases (because of higher consumption and lower conservation by users), effecting upward pressure on price. If a standardized futures market existed for most commodities, one could also observe the futures prices for them.

For convenience, we also assume that the correlations of commodity prices with the market are low; thus, they do not carry much systematic risk (β of zero). In most cases, we assume a real risk-free rate of zero. Thus, throughout this book, we assume that futures prices are unbiased estimates of future spot prices. None of these assumptions is required for analysis, but they help us understand the stylized case studies presented without a lot of complexity. Once the mechanics are fully understood, you can modify these assumptions if needed. We also use futures and forward prices interchangeably and ignore effects such as counterparty risks and liquidity premium in less standardized forward and swap contracts. In the risk-neutral framework, the stochastic price process for commodities will have a risk-neutral drift (in our case, the drift is zero), and the expected return on a futures contract is zero.

Exploration and Development Project for Natural Resources

Natural resources such as metals and crude oil require different stages of exploration and development (E&D). In each stage, investments are needed, and decisions may depend on various technical uncertainties, such as geological, technological, and market-based price uncertainties. Natural resources explorations share many common features with research and development (R&D) projects in life sciences.

First consider a copper mine. Several staged investments need to be advanced in discovering and exploring a site for copper. These investments can be stopped at the beginning of a stage if the price of the commodity combined with the expected technical feasibility of the mine indicates that abandonment is optimal. If all four exploration stages are successfully completed, the company enters the development stage in which additional investments are needed to set up the mine for extraction. Typically, the company would have received a leasehold on the land it wants to mine from the government. This leasehold is for a specific period of time and the longer it takes to explore and develop the mine, the less time is left on the leasehold for extraction and sale.

For this stylized example, we consider a four-stage exploration project. Once all exploration stages are completed, the company has the option to enter the development stage in which it has fixed expenses in equipment and personnel to set up the mine. If the company completes all stages (which are in sequence), it will own a mine with a certain amount of deposits and will have certain market value that is a function of the price and cost of extraction of the commodity (in this case copper). We assume a binary technical feasibility outcome in each of the exploration stages. That is, after having taken the needed investment in any of the exploratory stages, the company may be forced to abandon the mine due to technical infeasibility. It may also abandon the mine if the expected revenues from the mine are less than the costs of exploration, development, and extraction. The model is shown in Figure 11.1.

There are several modeling assumptions to note. First, we model the price of copper as a mean reverting stochastic function (since copper is a commodity). The price of copper P currently is $6,000 ($6K)/metric ton (all dollars are in thousands in the model in Figure 11.1). Using historical data, we also estimate a minimum and maximum bound for copper at $5K/ton and $9K/ton, respectively. By analyzing historical price movements, we also estimate that the half-life for the price of copper is 4 years. This is the time taken for prices to revert halfway back to the long-run mean after an excursion due to a shock or other factors. Figure 11.2 shows sample risk-neutral stochastic simulations for copper prices. As explained, we assume the real risk-free rate to be zero.

We also model the operating costs (cost of extraction, concentration, production, and logistics) as a single parameter EC. We estimate the current cost

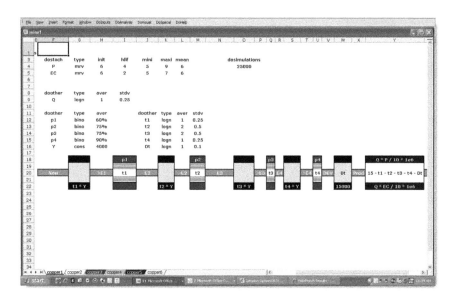

FIGURE 11.1
Model of the copper mine.

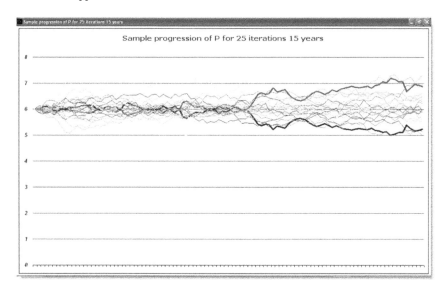

FIGURE 11.2
Stochastic simulation of copper prices.

of production and transportation per ton to be approximately the same as the price ($6K/ton). The range of operating costs is from $5K/ton to $7K/ton. Because equipment can be mothballed and returned to service, the operating costs revert to long-run averages more quickly than prices. We estimate the

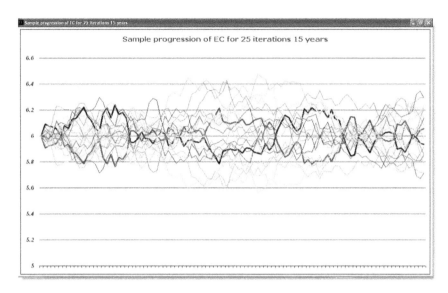

FIGURE 11.3
Stochastic simulation of extraction cost.

half-life of costs to be 2 years. The sample risk-neutral stochastic simulations of operating costs are shown in Figure 11.3. Compared to the price, the reversion rates are a lot stronger for costs because of the shorter half-life.

There is also a third variable for the total quantity of copper that can be extracted from the mine Q. We assume Q to be the quantity of final product (not copper ore), which is a combination of raw copper ore and the grade of copper. We consider the total quantity to be probabilistic. We model this as a lognormal distribution with an average of 1 million tons and a standard deviation of 0.25 million tons.

The four stages of exploration projects have binary technical success rates of 60%, 75%, 75%, and 90%, respectively. The time required to conduct these stages is probabilistic, with expected duration of 1, 2, 2, and 1 years, respectively. The stages have lognormal distributions with standard deviations of 0.25, 0.5, 0.5, and 0.25 per year, respectively. The costs of conducting these exploration stages are a function of time taken. We estimate this to be $4 million/year. For example, the cost of conducting stage II exploration is $t2 \times Y$, where $t2$ is the time taken, and Y is the cost per year ($4 million/year).

Once the four exploration stages are completed, the company will be ready to develop the mine. We estimate this to take $15 million to conduct, and the company has approximately 1 year (Dt) to commit to it. During this time, it negotiates with the equipment and logistics providers and prepares contracts to execute. One year after the commitment for development is made, the company receives the first revenues from the mine. The development costs are fixed costs (and sunk once taken).

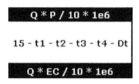

FIGURE 11.4
Model of the operating mine.

We model the functioning mine as a swaption (an option to swap cost for price every year) in which the operator observes both the price of copper and operating costs at the beginning of each year and makes a decision to produce that year or not. If the operating costs are lower, the operator will decide to shut down the mine for that year and try again next year. Price and operating costs are assumed to remain the same during the course of the year (only for convenience). Figure 11.4 is the model of the operating mine.

The total time of the swaption is the remaining time in the leasehold, which was set at 15 years at the beginning of the problem. The time for production is what is remaining after subtracting the time taken for E&D from the total duration of 15 years (represented in the middle). On top, we represent the revenue, which is a function of Q (total quantity of copper in the mine) and P (price/ton for copper. We assume 1/10 of Q can be extracted every year. This is multiplied by 1E6 (1 million) as we have represented Q in millions of tons. Similarly, the cost (at the bottom) is a function of the total quantity Q and extraction cost/ton EC. Every year, the operator has the right (but not an obligation) to swap cost for price by producing copper from the mine and selling it. Note that we ignore a mine shutdown, maintenance, and open costs for now. After the 15 years run out, the land has to be returned to the government, and no more production is possible.

Figure 11.5 indicates the risk-neutral payoff from the leasehold to explore and develop the mine. We value the mine at $23 million.

Figure 11.6 shows the cumulative probability distribution of the risk-neutral payoff from the leasehold. Note that there is only a 15% chance of making money, largely driven by the binary technical failure due to geology or some other technical issue, including lack of quality and complex soil characteristics. On the upside, there is a 10% probability that the mine will be worth over $69 million in economic value by entering into a contract for the leasehold today.

Now, consider a situation in which the mine is operated continuously. This may be due to technical reasons, costs of shutting down and starting up, or simply contractual constraints in the leasehold when the government does not let the company reduce employment. We can value the leasehold when such a constraint exists; the result is given in Figure 11.7. We find the value drops to about $4 million from the original $23 million. The difference of $19 million comes from the lack of operating flexibility to operate the mine only when revenue is higher than costs.

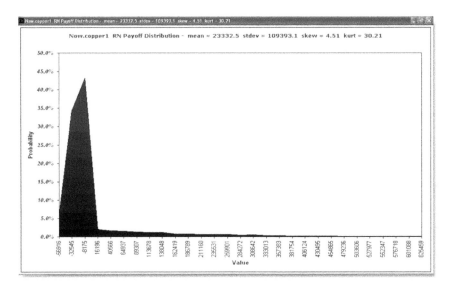

FIGURE 11.5
Risk-neutral payoff from the copper mine.

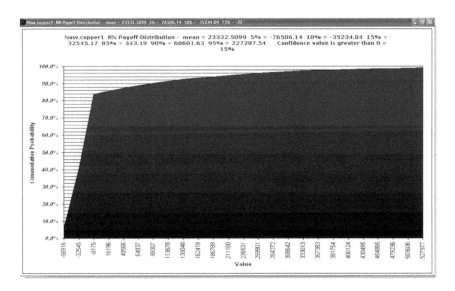

FIGURE 11.6
Cumulative probability of risk-neutral payoff from the mine.

We can also calculate the value of the mine if the decisions to go forward in the four exploration stages and the development stage are not options. The company is obligated to perform (perhaps by contract). This is also what we assume while preparing a discounted cash flow (DCF) analysis. As shown

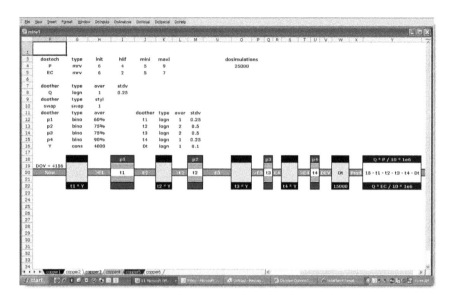

FIGURE 11.7
Model of a continually operating mine.

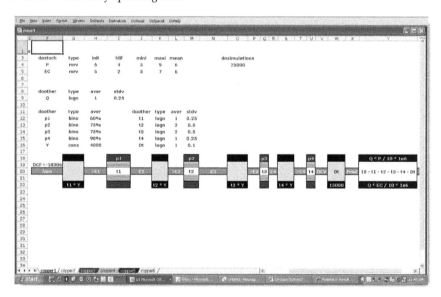

FIGURE 11.8
Model of the copper mine with no flexibility.

in Figure 11.8, if we remove all flexibility (ability to terminate E&D as well as the operating flexibility of the mine), we get –$18 million, indicating a negative net present value (NPV) project. In this case, the simple DCF analysis will provide a "no investment" decision for the company, and if the company

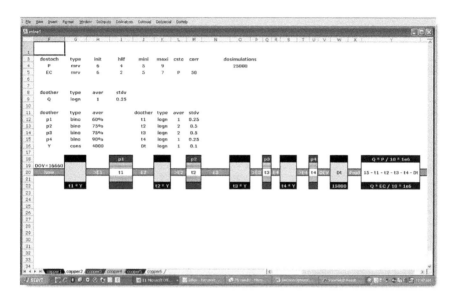

FIGURE 11.9
Model of the copper mine with extraction costs correlated with price of copper.

follows that prescription, it will walk away from an investment proposition worth over $24 million.

Also consider a situation in which the operating cost of the mine is correlated with the price of copper. This is a realistic scenario because as demand for copper increases (and its price increases), the owners of mining equipment will raise their rates, and labor will demand higher wages. In the model in Figure 11.9, we assume that the operating cost is correlated with price of copper. In this case, we find that the value of the leasehold decreases to approximately $17 million from the original $23 million.

Let us add a "speeding up" option for the development phase. After completing the exploration stages, we initially assumed that the company can commit $15 million to develop the mine, and it has approximately 1 year to commit to that development. Two years after the exploration stages are completed, the company receives the first revenue from the mine. Now assume that there is a speeding up option it can consider. By spending more money ($25 million), the company can reach the first revenue stream in 6 months (instead of the 1 year initially assumed). This is an option for the company, and it does not necessarily need to take that route. It may chose to do this if the increased time of production coupled with the price and cost levels justify the additional investment. The model with the additional option and results is given in Figure 11.10. The additional option increases the value of the leasehold by an additional $4 million to a total of $27 million.

These types of acceleration options could be very important in bidding for the leasehold. If the company has a better estimate of the ultimate value that

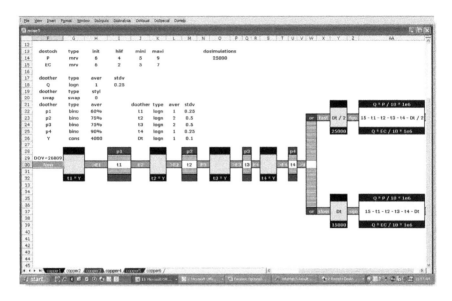

FIGURE 11.10
Model of the copper mine with option to speed up development.

can be extracted from the leasehold, it will have an advantage bidding for the opportunity to develop the mine.

In all the previous examples, we considered the price of copper as mean reverting, and the only market information we can see is the current spot price. In many commodity markets, there are well-functioning futures that allow us to see the market's expectation of future prices. We consider the futures price to be an unbiased estimator of future spot prices. Not taking this information into account can introduce significant errors into the valuation of the leasehold. The forward or futures curve aggregates the expectation of a large number of market participants who have many different types of information. It also shows the equilibrium level of futures prices between hedgers and speculators. This is crucial information to be considered while valuing an asset that critically depends on future commodity prices.

One way to incorporate the futures prices into our analysis is to let the mean reverting price process to revert to the futures price rather than the long-run mean. Futures prices may only be available for a few years and may not cover the entire time span of the problem considered. For the current problem, the horizon is 15 years. Assume that we can observe futures prices for the first 5 years. After the first 5 years, we let the futures curve back to the long-run mean. Figure 11.11 is the representation of this information.

Futures data show falling copper prices for the first 5 years (for convenience, we represented this in yearly numbers). After that, no specific information is

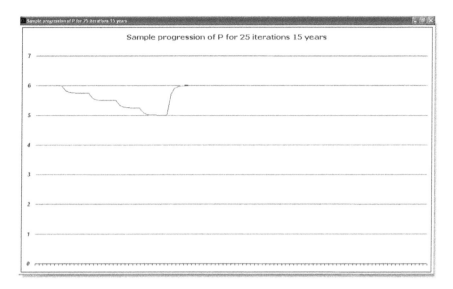

FIGURE 11.11
Futures prices of copper.

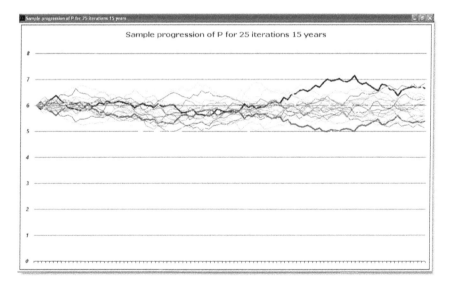

FIGURE 11.12
Stochastic simulation of copper prices with futures information incorporated.

available, so we let the prices revert to the long-run mean of $6K/ton. Sample risk-neutral simulations of copper prices with the 5-year futures price information are given in Figure 11.12. Note that the futures information does not guarantee that prices will be exactly as implied by the futures prices.

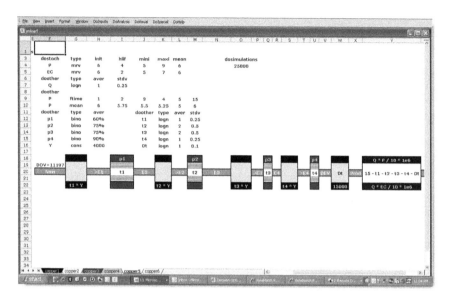

FIGURE 11.13
Model of the mine with all pricing information incorporated.

Figure 11.13 shows the value of the leasehold with the incorporation of the futures information. Note that the value of the leasehold drops appreciably to $11 million from the previously calculated $23 million. Futures information cannot be ignored in this type of analysis.

Governments provide leaseholds to commercial companies using a bidding process both to develop the mine (which reverts back to the government at the end of the leasehold) and to create local employment. If the leaseholder decides not to develop the land and simply lets the leasehold expire, the deal does not serve the purpose. It is unlikely that the commercial company will explore or develop the land if it is not going to make a profit. In some situations, the government may provide incentives to incite E&D. The higher the copper prices are, the less likely that the company will require an incentive to undertake the project.

Let us analyze an incentive scheme in which the government pays the company some amount per year if the price of copper at the beginning of the exploration or the development phase is less than $7K/ton. The amount is calculated as ($7,000 − price of copper) × $1 million. This will reduce the cost of E&D for the company when copper prices are low (situations that may result in the abandonment of the mine project). The lower the price of copper is, the higher the subsidy from the government will be. Figure 11.14 provides the (partial) model showing how the subsidy is modeled in each year. The value of the leasehold is increased to $28 million from the original $23 million due to the government subsidy. This also reduces the probability of abandonment.

FIGURE 11.14
Model of the mine with government assistance in exploration phases.

Let us summarize the results from the various analyses:

Decision Options value of the leasehold as originally modeled	$23 million
Value when the mine is operated continuously once developed	$4 million
Value when operating costs are correlated with copper price	$17 million
Value when all flexibility is removed (DCF value)	−$18 million
Value when a speeding up option exists in development	$27 million
Value when the futures information is incorporated	$11 million
Value when the government subsidy is included	$28 million

All of these are different problems with added features or constraints to the original problem. Note that the DCF value is the same as when no decision or operating flexibility exists, and it shows a negative NPV, implying the leasehold is not a worthwhile investment. Intuitively, that does not make sense as the leasehold, as an option to explore and develop, cannot be worth less than zero.

If an options-based analysis cannot be performed, decision makers may use historical proxies in the bidding process. As we have seen, information such as the market-based futures prices (which incorporate all known information) and correlation of costs to prices significantly affects the overall value of the mine. Using historical proxies could lead to erroneous values and incorrect bids. Winner's regret happens in these bidding contests as the winner of the leasehold realizes that it has overbid and thus overpaid for the asset.

Hydrocarbon Exploration, Development, Production, and Risk Sharing

Hydrocarbon exploration, development, and production are extremely expensive and involve long processes that carry significant private and market risks. Although the visible tactical profits of oil companies become the target of public anger and political sound bites, in times of high oil prices, the risks taken and investments needed to explore, develop, and produce from an oil field are often forgotten. The length and risk of this E&D process of hydrocarbons share some common characteristics with pharmaceutical R&D, an equally misunderstood process by the public in general and policy makers specifically.

Large oil companies that hold leases and other rights on hydrocarbon-rich areas often rely on the expertise and technology of oil service and equipment companies during the E&D phase. The U.S. oil and gas services and equipment industries include over 8,000 companies with a combined total annual revenue exceeding $25 billion. Large global suppliers include Halliburton, Schlumberger, and Baker Hughes. This industry is highly fragmented, and the small specialty firms with expertise in specific aspects of E&D are the norm. Demands for E&D and production services and equipment are normally driven by the price of oil and gas. Large companies such as Schlumberger can provide the entire spectrum of services demanded by the producers, and small firms usually focus on specific aspects of the process. Over one-third of the revenue for oil service and equipment companies comes from drilling services as well as drilling equipment. Nearly half the revenue comes from preparing wells for production, maintenance/ enhancement of producing wells, and exploration. Equipment such as offshore drilling rigs require long lead times for design and manufacture and have complex supply-and-demand dynamics.

As the world hunger for oil remains unabated, over 82 million barrels of "black gold" are consumed around the world every day. The United States accounts for 25%, and emerging BRIC accounts for 20%. Currently, the United States, BRIC, Japan, and Germany soak up over 50% of the world's supply of oil. With countries of BRIC showing high growth rates, it is possible that consumption in these four countries can dwarf the rest of the world in a decade. With progress in alternative energy stagnating, world growth and stability still depend significantly on the identification, extraction, and conservation of the energy-giving hydrocarbons.

Oil exploration is an expensive and high-risk operation. Typically, only governments or large oil companies embark on such endeavors. With oil prices showing high levels of volatility (driven by demand, supply, storage, and perception of remaining stock) and the known deposits of the world's oil shrinking, oil companies find it increasingly difficult to make optimal decisions in E&D. This is a problem faced by the executives of GiantOil.

GiantOil has E&D rights for a large area in western Africa. It won the E&D rights in a bidding round that concluded a couple of years ago and for which it paid handsomely. Although the timing appeared right for the creation of a profitable operation there, GiantOil executives have been hesitant as they felt that the current price level of oil (at \$100/barrel) was unsustainable. With long lead times involved in E&D, their gut feeling was that initiating the process now will put them at the risk of production commencing when prices normalize (driven by mean reversion in oil prices). Some of the GiantOil executives have been around for a while, and they have had bad experiences in the past by getting excited when prices peaked and commissioning large projects at significant expense only to find the price of oil, profitability, and the stock price fell when they were ready to produce.

ClearTech is an oil services and equipment company that is considered highly innovative. ClearTech's management has been thinking of ways to enhance shareholder value through more innovative business processes in addition to their ongoing R&D. During a recent executive management team meeting, the chief financial officer (CFO) discussed an idea that found some traction. She suggested ClearTech share the risk of exploration, development, and production with GiantOil. GiantOil has been a long-term customer for ClearTech, and the management teams at both companies have a long history of collaborating on projects and solving problems together. ClearTech management knew that GiantOil has been hesitant to initiate work in western Africa because of the timing risk and the large investments needed. If ClearTech can share some of the risk, it may induce GiantOil to initiate the investment. Such collaboration also gives ClearTech a direct path to getting involved in E&D in the new area. The project will be large, and it will ensure a steady stream of activity for ClearTech. However, this was new territory for ClearTech since its contracts with customers thus far had been conventional, prescribing fees for services and timelines. Sharing risk will mean that it must settle for lower revenue up front for possible gains in the future. In this case, both the private risks (risk of not finding the anticipated quantity and quality) and market risks (risk of lower oil prices and higher production costs by the time production commences) have to be shared with the customer. ClearTech will have to invest into the E&D process along with GiantOil, and if the expectations do not materialize, their stock prices will take a hit. The CFO was confident that they could evaluate and price the risk. She felt that ClearTech may have some advantages in pricing a complex contract with GiantOil.

In the first meeting between ClearTech and GiantOil, they discussed some ideas both companies could consider in constructing a contract:

1. GiantOil will fund some of the E&D costs. It would like to keep these as a fixed amount and let ClearTech pick up the remaining cost. GiantOil felt that the cost of equipment and services are within the control of ClearTech, so such an arrangement will provide incentives for ClearTech to manage E&D costs to a minimum.

2. In the exploration phase, GiantOil will share the cost of exploratory wells but not of the detailed seismic survey. A preliminary seismic survey has already been done, and GiantOil will make the results available for ClearTech. If ClearTech feels that the preliminary seismic survey results are not sufficient, it can embark on a more detailed survey. Since GiantOil will not fund a detailed survey, ClearTech would be responsible for all those costs. ClearTech has multiple ways to execute the project; it can go ahead with exploratory drilling first (with costs shared by GiantOil), or deploy its own detailed seismic survey technology at its own expense to assess the probability of success before embarking on exploratory drilling. ClearTech can abandon the project if the seismic survey results are not good enough to go further. If the exploratory drilling returns dry holes, it can then abandon the entire project as well.

3. In the development phase, GiantOil will share the cost of development and injection wells but not any technology deployed by ClearTech to assess the size and quality of the hydrocarbon deposit. As in the exploratory phase, ClearTech has multiple ways to execute the project; it can go ahead with full-fledged development of the field (with costs shared by GiantOil), or deploy its technologies to obtain a better assessment of the nature, number, and depth of the development wells to be drilled for optimal production. The new ClearTech technology uses electromagnetic inductance to have a better assessment of the oil field. ClearTech can abandon the project at this stage as well if the assessment indicates that the continuation of the project may be unprofitable.

4. GiantOil profits have to be at some level for it to consider initiating the project. If prices fall below $75/barrel, it will become extremely difficult to manage it. GiantOil wants to share profit with ClearTech as a function of oil price. It will share a larger percentage of the revenue if the price is above $75/barrel. The revenue share will be at a lower bracket if prices drop below $75/barrel.

The CFO assembled a group of experts to gather the necessary information. They needed to get an idea of what terms will be economically viable for ClearTech to consider and estimated the following parameters for the project:

Cost of an exploratory well	$8–$15 million, $10 million mode
Number of exploratory wells needed prior to seismic survey	10
Timeline for exploration	0.5–2 years
Probability of success in exploration	60%
Cost of seismic survey	$40 million
Probability of success in seismic survey	80%

Time for seismic survey	6 months
Number of exploratory wells needed after seismic survey	6
Chance of success after exploratory wells	75%
Cost of development	(1,000/quality) million
Range of quality	0.5–1.5, 1.0 expected
Cost of quality assessment technology	$100 million
Cost of development after quality assessment	(800/quality) million
Cost of production per barrel	20/quality/barrel
Rate of production per year	25–40 million barrels/year, 30 million barrels/year expected

In the second meeting, GiantOil proposed that ClearTech be responsible for all costs (except the reimbursement rates discussed in the following paragraphs). ClearTech also has the option to walk away from the oil field at specific points in the decision process. These options to abandon the project come at the following decision points:

After the seismic survey

After the exploratory wells are completed and data are available

After the technology implementation to assess the quality of reserves

After the development is completed and the field is ready for production

GiantOil will also reimburse ClearTech in the following fashion:

Fixed costs reimbursed for exploratory wells	$5 million/well
Fixed cost reimbursed for development	$250 million

This means that if the exploratory wells cost more than $5 million, the cost will come from ClearTech's pockets. In essence, ClearTech will be investing along with GiantOil. Similarly, the cost of development less the fixed reimbursement of $250 million will be invested by ClearTech. These are ClearTech's options, and it can always walk away from further investments at the decision points described. GiantOil felt that this arrangement in which ClearTech is responsible for the total costs of seismic survey, technology implementation, exploratory well drilling, and development preparation will provide incentives to be as efficient as possible. From GiantOil's perspective, it will not run into any cost overages as the payments during the E&D phase are fixed. ClearTech is thus assuming all the cost overrun risks of the project. In the production phase, ClearTech will be responsible for production costs of oil over and above $5/barrel. Here again, GiantOil would like to eliminate cost overruns because it will be contracting with ClearTech for the production phase as well. For example, if the cost of production is $20/barrel, ClearTech will be responsible for $20 − $5 = $15/barrel cost of production.

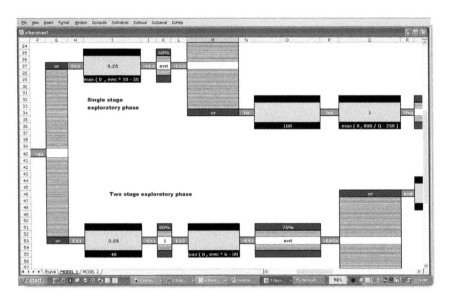

FIGURE 11.15
Model for the deal showing exploratory phases.

In return, GiantOil will share revenues from the field once it starts to produce. GiantOil will pay ClearTech 33% of the revenue for oil prices up to $75/barrel and 75% of the revenue for prices exceeding $75/barrel. For example, if the price of oil is $90/barrel and the production was 30 million barrels a year, the revenue attributable to ClearTech will be calculated as follows:

$$30 \text{ million} \times (75/3 + (90 - 75)*3/4) = 30 \times (25 + 11.25) = \$1087.5 \text{ million}$$

Clearly, the higher the oil prices, the better for ClearTech. GiantOil's management reduced the downside risk of the investment significantly but also provided some of the upside potential to ClearTech to entice it into the arrangement. The first thing the ClearTech CFO wanted to do was to assess the economic gain from the proposed deal with GiantOil. Since there are many uncertainties (such as oil price, quality, timelines, and costs) as well as decision flexibility, the CFO decided to create a decision options model of the deal. The partial model is given in Figure 11.15.

The decision options model shows the decisions, choices, and private risks. For example, at the onset ClearTech managers can decide if they want to go ahead and commission the 10 exploratory wells needed to determine if it is possible to develop the field. GiantOil will pay ClearTech $5 million/well, so the investment GiantOil must commit is $ewc \times 10 - \$50$ million, where ewc is the cost of the exploratory well. After this investment is taken, there is only a 60% chance that ClearTech will find a feasible development opportunity. On the other hand, it can commit $40 million for a detailed seismic survey of the area. Note that ClearTech is alone here as GiantOil will not contribute to the seismic survey. ClearTech estimates that there is a 20% chance that it will find the field is not viable after the seismic survey, in which case it will abandon the entire project. If it obtains data that show viability, it can go ahead and conduct exploratory drilling. With the seismic data, it can better position the wells, so only six wells are needed to reach a conclusion of feasibility. For those six wells, GiantOil will pay $30 million ($5 million/well). The data from these wells will give ClearTech a 75% chance of going ahead. Note that the technical probability of success for the oil field is still 60% (80% × 75%), but in the latter case ClearTech obtains information in two bundles and can split the investment decisions, providing higher flexibility. However, this delays development by 1 year, and ClearTech is responsible for a higher percentage of the total cost and has some expectations of the time needed to conduct the exploratory phase and a better handle on the seismic survey timeline (they estimate 1 year), but the exploratory drilling phase may take 6 months to 2 years depending on availability of equipment and qualified personnel.

Once the exploratory phase is over and the data show that the development is feasible, then the development phase can begin (Figure 11.16). Here again, ClearTech has multiple choices. Even if the exploration technically succeeds, it can decide not to develop if the economics are not attractive (maybe due to falling oil prices, rising costs, or both). It can enter full-fledged development with GiantOil footing $250 million of the total development cost. The cost of development is (1,000/quality) million, where quality is a variable between

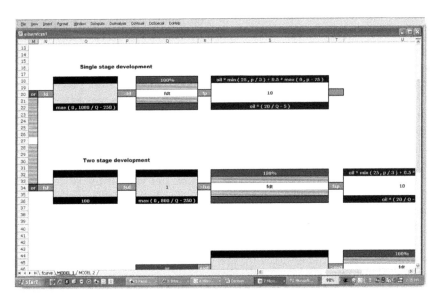

FIGURE 11.16
Model for the deal showing development phases.

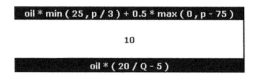

FIGURE 11.17
Details of production.

dostoch	type	init	vola	hlif	mean	jpro	jmag	jhli	jvol
p	mrl	100	20	3	50	2	50	2	30

FIGURE 11.18
Stochastic characteristics of oil price.

0.5 and 1.5. It is a proxy for the geological characteristics of the basin. This means that the cost of development ranges from $667 to $2,000 million. It has substantial uncertainty. Alternately, ClearTech can commission an implementation of a new technology it owns for more information on the quality, which will allow it to design a better configuration of wells and injection systems. It will cost ClearTech $100 million as GiantOil does not share these costs. The $100 million in investment in technology reduces the cost of development to (800/quality) million for ClearTech. So, the cost of development reduces to a range of $533 to $1,600 million, primarily because the new information allows a more efficient design. However, the technology phase will extend the timeline by 1 year.

Once the development phase is completed, production can start (Figure 11.17). The companies estimate a production of 25 to 40 million barrels/year, with an expected production of 30 million barrels. The cost of production is also related to the quality of the find. They estimate a production cost of 20/quality per barrel. This means that the cost of production is in the range of $13.3 to $40/barrel. GiantOil pays $5/barrel of the production cost, so the cost of production for ClearTech is in the range of $8.33 to $35/barrel. The revenue shared with ClearTech is a function of the price of oil. Revenue is represented in the equation on top, and cost is represented in the equation at the bottom. The symbol p represents price of oil, Q represents quality, and oil is the total production per year. If the price of oil is below $75/barrel, ClearTech gets only one-third of the revenue stream. If the price is above $75/barrel, ClearTech receives 75% of the revenue over and above the $75 level.

The negotiating team also analyzed recent oil prices and created the parameters shown in Figure 11.18. The current price of oil is high, $100/barrel compared to the historical long-run average price of $50/barrel. Oil prices, just as the price of other commodities, is mean reverting. They estimated an annual volatility of 20% and a half-life of 3 years along with a 2% chance that in any year the oil price takes a jump (a shock) from normal. When such a

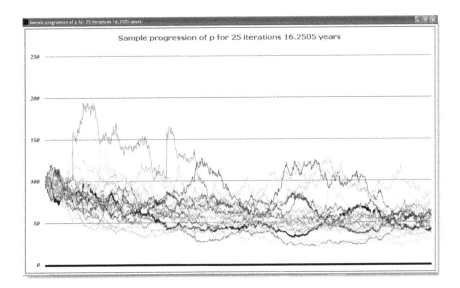

FIGURE 11.19
Stochastic simulation of oil price.

shock happens, the price can jump by 1.5 (50% more than the price at which the shock happens). After the jump, mean reversion increases with a corresponding decrease in half-life to 2 years. The volatility also changes to 30% from the normal level of 20%.

Mean reversion in prices tends to pull prices back even after severe shocks to the system. Positive shocks will make many of the marginal oil fields in the world viable, and the supply will increase as producers scramble to pump more. Meanwhile, consumers and industrial users will pull back on consumption. As new supply comes on line and demand drops, driven by higher prices (and possibly slowing economy due to the shock), an oil glut results, and prices normalize. The reverse happens in the case of negative shocks as many fields shut down, and consumers shelve all desire to conserve. Figure 11.19 shows sample simulations of the oil prices with these assumptions.

The team also modeled a variable to represent quality of the geological characteristics. They modeled this as probabilistic. They estimated these parameters to be in the range of 0.5–1.5, with an expectation of 1.0. So, they can have surprises both on the positive and negative sides.

Similarly, the production rate is a function of total quantity as well as the configuration of wells. They expect to produce 30 million barrels/day, but production is in the range of 25 to 40 million and depends on a variety of factors. They modeled this as probabilistic as well.

With the assumptions on the characteristics of the oil field and suggested contract terms by GiantOil, ClearTech's team set out to assess the shareholder value, if any, accruing to ClearTech if it were to enter the deal. Figure 11.20 indicates the risk-neutral payoff distribution to ClearTech from the contract.

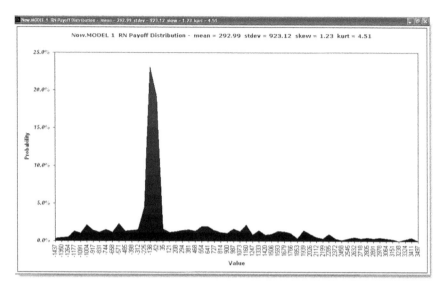

FIGURE 11.20
Risk-neutral payoff from the deal to ClearTech.

The results show that the value accruing to ClearTech is approximately $300 million. The peak represents abandonment either after seismic survey or after exploratory wells, either for technical reasons (most likely) or for market reasons. There is also a small probability of abandonment in development as represented by the negative side of the distribution. This is related to conscious decisions made by ClearTech to abandon the field and walk away from the contract because the economics have deteriorated. This may be due to the fact that oil production is not viable after development due to a collapse in oil prices, an increase in production costs, or both, leading to nonviable production economics. It is assumed that the oil field will be functional for 10 years, and ClearTech can make production decisions in any year depending on price of oil and costs.

Figure 11.21 shows the cumulative probability of risk-neutral payoff for ClearTech. It shows that ClearTech has a 35% confidence in making money from the contract. There is significant risk, but overall it is of significant value to ClearTech.

One piece of information it has not taken into account in the current analysis is the forward price of oil. Since oil is a well-traded commodity around the world, the market has certain expectations for its price in the future. These can be seen in the forward curve of oil prices. Figure 11.22 shows the forward curve, and Figure 11.23 shows the sample of simulated prices with the forward curve information incorporated in it.

Reanalyzing the contract with forward information incorporated, ClearTech found the value of the deal to be close to $287 million, roughly the same as the earlier value.

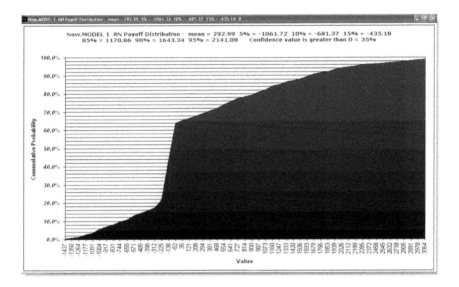

FIGURE 11.21
Cumulative probability of risk-neutral payoff to ClearTech.

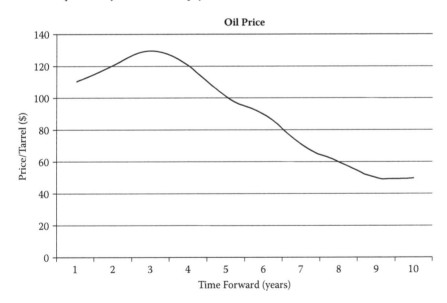

FIGURE 11.22
Forward curve for oil price.

Armed with the information that the deal adds approximately $300 million in shareholder value, ClearTech's management wanted to accept it. However, the CFO felt that GiantOil may have set the terms based more on experience than on analysis. ClearTech may have an opportunity to renegotiate some

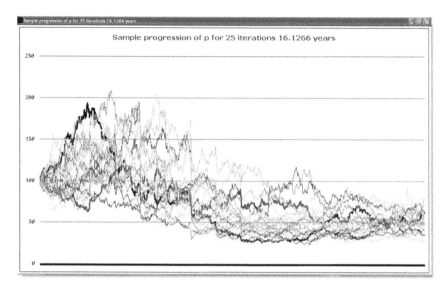

FIGURE 11.23
Stochastic simulation of oil price with forward curve information.

of the deal terms to its advantage. This means determining what aspects of the contract were the most important to GiantOil. In conversations with the analysts, ClearTech understood that the up-front costs were more important and wanted to reduce the cost of the exploratory well from the current $5 million/well and in return reimburse a higher percentage of the production costs. Based on this information, ClearTech managers wanted to suggest an alternative deal to GiantOil. ClearTech would pay all of the exploratory costs; that is, there will not be any reimbursement of exploratory well costs by GiantOil. In return, though, ClearTech would like GiantOil to pick up more of the production cost per barrel of oil. Instead of the current $5/barrel reimbursement, ClearTech wants $8.5/barrel. This suggestion was welcome news for GiantOil's management, who sought to further reduce its risk.

Using the new contract terms, ClearTech reevaluated the deal and found that the value of the contract to them was enhanced to approximately $1 billion. The ability to price and buy risk allowed it to negotiate a contract of significantly higher value (more than three times the value of the original contract). GiantOil's aversion to risk in E&P cost it significantly (an incremental $700 million in value was transferred from GiantOil's shareholders to ClearTech shareholders).

The CFO also thought of other creative ways to increase value by buying a put option to sell the oil field back to GiantOil under certain conditions. Note that ClearTech revenue and cost streams are different from those of GiantOil, so even if it is unprofitable for ClearTech to operate the oil field, it may still profitable for GiantOil to do so. For example, the CFO thought of including an option to sell ClearTech's rights to produce from the field after the development is completed for a price of $3 billion.

From GiantOil's perspective, this was another good outcome. The expected output of 30 million barrels/year for 10 years put the total production from the patch at 300 million barrels. The $3 billion price tag implies a profit per barrel of $10. With expected cost of production of $20/barrel and the long-run expected price of oil at $50/barrel, the probability of less than $10/barrel profit is small. GiantOil management jumped at the chance of adding this additional feature in the contract. From ClearTech's perspective, it adds more flexibility because it is an option given (at GiantOil's expense) and it will only exercise the option if doing so is profitable (compared to operating the field under the contract conditions). If the cost of production is high and prices are low, it may be worthwhile to exercise the option to "put" the field for $3 billion and walk away. This actually caps the downside risk for ClearTech. Including such a put option, the ClearTech team calculated the value of the deal again and found that to be higher.

Design of Subsidies and Incentives for Alternative Energy Technologies

Governments around the world have been struggling to create policies that will aid the long-term strategic development of their countries. The recent "superspike" in crude oil prices has understandably sent shock waves through the entire world as many countries struggle to avoid recession (or even stagflation), and some others are unable to find the "energy" to feed their hypergrowth. My goal here is the demonstration of the application of decision options in policy making and not to influence policy. All examples here are stylized, and the numbers used are only for illustration.

Ethanol is a substitute for gasoline (petrol) for automobiles and can be produced from plant materials such as corn. Ethanol production in the United States has risen from a paltry 175 million gallons in 1980 to close to 7.5 billion gallons in 2008. The United States leads in ethanol production using corn, closely followed by Brazil, which uses sugarcane as the input to production. Worldwide production is estimated to be 15 billion gallons per year, approximately twice that of the U.S. production. The latest readings show over 150 ethanol plants in the United States. Subsidies and incentives are a big part of ethanol production in the United States.

Ethanol is typically produced by the dry mill process. First, the corn kernel is ground into flour. It is then mixed with water and enzyme to form a slurry. The slurry is then heated to reduce viscosity before it is pumped into a pressurized cooker and cooked for 5 minutes. The mixture is then cooled and held in a tank for couple of hours. During this time, the starch in the flour is broken down by the added enzymes. After this, another type of enzyme is added, and the mixture is pumped into fermentation tanks. Yeast is then added to convert the mixture into ethanol and carbon dioxide. The fermentation process takes approximately 3 days. The mixture from the fermentation process contains 15% ethanol. It is then distilled to boil the water off and create a 95% ethanol mixture. A dehydration process follows to remove the remaining water and produce 100% ethanol, which is the end product. It is then sent to storage tanks or directly to the terminal for distribution. This process produces two by-products, distiller's grains and carbon dioxide. The mixture from the distillation tanks (where ethanol is purified) contains grain, yeast, and water. Using centrifuges, this can be converted into "wet distiller's grain," which is excellent cattle feed. The fermentation process also produces large amounts of carbon dioxide that can be captured and purified by a scrubber and marketed to food processors for use in carbonated beverages or for dry ice (frozen carbon dioxide).

Consider the decision to build an ethanol plant of 50 million gallons/year capacity. We assume actual production per month to be 4–4.5 million gallons. Assume a capital cost of $80–$140 million to be invested equally in four stages; each stage is 6 months long. The construction costs are uncertain due to a variety of factors, including land cost, equipment and construction labor, weather, and unanticipated outages.

The inputs into the process are primarily corn, water, enzymes, and other chemicals. The variable costs include electricity, other fuels, labor, maintenance, and other administrative costs. The cost of production and the price of output have the following components:

Fixed operating cost for personnel and maintenance	$2 million/year
Cost of inputs (corn, enzymes and other additives)	$1.70–$2.50/gallon
Cost of production (electricity and other variable costs)	$0.45–$0.60/gallon
Cost of transportation, storage, insurance, and other	$0.50–$0.80/gallon
Price of ethanol at terminal (same as automobile gas/petrol)	$2.50–$4.00/gallon
Price of by-product 1 (dried grain)	$0.20–$0.30/gallon
Price of by-product 2 (carbon dioxide for bottlers and dry ice makers)	$0.03–$0.05/gallon
Life span of equipment and facility	8–15 years

We assume that the real risk-free interest rate is zero.

Let us calculate the value of the ethanol plant as a fixed plant where production happens regardless of prices of outputs and costs of inputs. We represent all prices of outputs and costs of inputs as mean reverting processes as all of them are commodities. By analyzing past price behavior, we can also calculate the rate of mean reversion (or half-life) for each of these items. We represent the cost of construction as mean reverting, and the components of cost (such as equipment and labor) are driven by supply and demand. By analyzing the historical equipment leasing and labor rates, a reversion rate can be calculated. The model and results are shown in Figure 11.24.

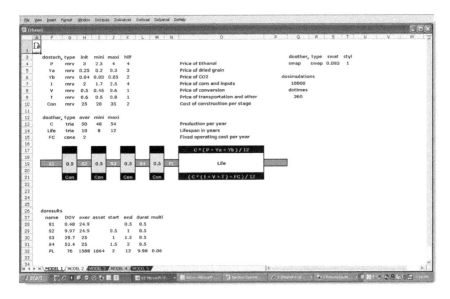

FIGURE 11.24
Model of the ethanol plant including staged investments.

All prices—ethanol, dried grain, and carbon dioxide as well as all input costs (corn, conversion, and transportation)—are modeled as mean reverting with appropriate parameters calculated from historical data. These are stochastic and can be observed over time. The capacity and the life of the plant are modeled as probabilistic. The fixed cost is a constant $2 million per year. The cost of construction in every stage is also stochastic. The costs for each stage are fully correlated. The higher the cost of construction in one stage, the higher the costs in subsequent stages.

The results show that the value of the plant is nearly zero today. The expected gain from the plant of $76 million and the flexibility value of the four-stage construction phase (which cost a total of $100 million) nearly cancel each other.

If the plant is run continuously, it will be worth $76 million (after construction). In this case, we are assuming that the plant will operate at a loss in some months if the revenue from the outputs (ethanol and by-products) is lower than the cost of production (inputs, conversion, and transportation). Consideration of the staged construction options (which allow abandonment of the plant at any time) provides an NPV that is close to zero. No investor is likely to build an ethanol plant if he or she assumes that the plant will be operational all through its life.

Now, consider the situation in which the plant can be operated only if the revenue is greater than expected cost. The plant operating decision will be made every month (0.0825 years in the model). Every month, the plant manager will observe the prices for ethanol and all by-products, compare those

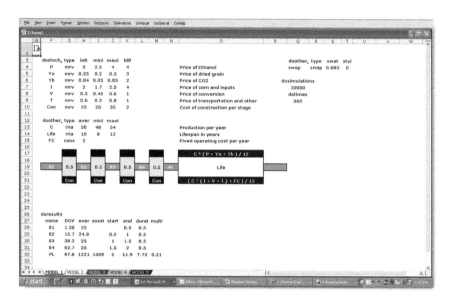

FIGURE 11.25
Model of the ethanol plant with flexibility to shut down and start-up.

with all the costs (input, conversion, and transportation), and make a decision whether to operate the plant. If the revenue is higher than costs, the manager will operate the plant. Otherwise, the plant manager will shut it down or, if it is already shut down from a previous month, keep it shut down. We assume that there are no additional costs to starting or shutting down a plant. The results are presented in Figure 11.25.

In this case, we find that the plant will be operational approximately 78% of its life span (when revenues exceed cost), and that the plant is worth an expected $88 million after construction. By keeping the plant idle about 22% of the time (when costs are high or revenue is low), the plant value can be increased from $76 to $88 million. The value of investment at the start of construction is still very low ($1 million). The reason it has a positive NPV is due to the staged investment that allows 2 years of observing the market and possible abandonment at any stage.

The numbers used in this example are roughly correct. You may be wondering what is causing the enormous jump in ethanol production in the United States. The answer is "subsidies." For example, suppose there is a subsidy from the government to induce investment in ethanol. There are a number of ways this subsidy can be provided. One way is to provide money to the producer when it sells ethanol. Consider a blending subsidy of $1 per gallon. This subsidy is available to any company that takes ethanol, blends it with automobile gasoline, and sells it. Since ethanol producers have the option to extract this subsidy by blending, most producers will also attempt to forward integrate into blending from production.

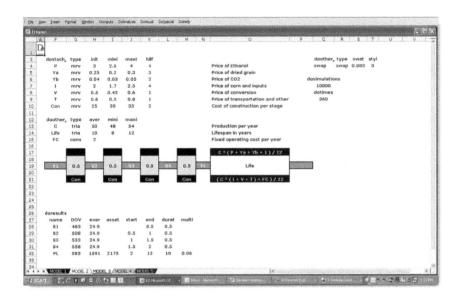

FIGURE 11.26
Model of the ethanol plant with blending subsidies captured by the producer.

Figure 11.26 shows the model and result with a blending subsidy of $1/gallon (half from the federal government and the other half from various state government schemes). The following items are noteworthy:

1. The value of the plant has increased significantly from $1 million to $483 million. This is essentially a direct transfer of government funds to the operator to entice the construction of the plant and production of ethanol. The total subsidy provided by the government is $500 million (approximately $1 × 50 million gallons × 10 years).

2. The plant is operational 100% of the time (instead of the original 78%) as the subsidy has made it profitable virtually all the time.

3. In effect, this is a direct transfer from the taxpayer's pockets to the ethanol manufacturer. The transfer of $500 million resulted in a $483 million gain for the manufacturer, resulting in an overall economic loss of $17 million.

From a policy perspective, one has to wonder if this is an optimal one. The policy has effected investment in ethanol production, but the production now has no market-based decisions. Many plants will be built, and production will be continuous. Policy makers bought this outcome by a straight transfer of $500 million to the ethanol manufacturers. The policy has not only distorted allocation of capital to efficient production methods but also resulted in a net economic loss to society (not to mention the taxpayers, who may have better use for the $500 million). If the policy makers really want to save the world by producing ethanol, there may be better policies.

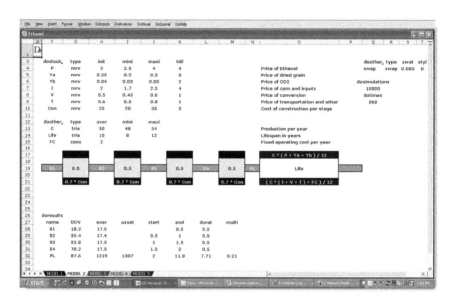

FIGURE 11.27
Model of the ethanol plant with construction assistance.

Now, consider another type of subsidy. To induce investment, the government can provide construction assistance. For example, it could offer tax breaks during construction. Assume that the net tax break for the investor from such a scheme is 30% of the investment. The cost to the government is $30 million (30% of the expected $100 million to be invested). The economic value of the plant under such a scenario is given in Figure 11.27. We reduce the investment needed at each stage by 30%.

In this case, the value of the plant increases to $18 million from the original $1 million. This is significant and will induce investment. The value of the plant after construction remains $88 million. The $70 million investment returns a plant worth $88 million in 2 years. This shows an internal rate of return of approximately 16%, attractive enough to many classes of capital providers as the private risk in the investment is very little as the ethanol production is a mature process (already applied by hundreds of plants across the world). The government can induce investment in ethanol (if that is the right thing to do) by providing incentive during the construction phase (at the fraction of the original subsidy cost).

If the policy makers are set on the blending subsidy, they could try to improve it by providing a subsidy only when they want to induce production. For example, from a societal standpoint, production is suboptimal when cost of inputs (which include corn, which is a feedstock) is higher than the price of ethanol. In this case, the act of inducing production through blind subsidies creates suboptimal use of resources and may cause unintended consequences such as grain price inflation. When the cost of inputs

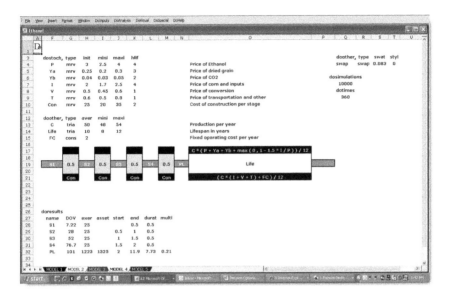

FIGURE 11.28
Model of the ethanol plant with varying subsidies based on price of output and cost of inputs.

is low, it is good to convert excess grain into energy. Consider a subsidy as given next:

$$\text{Subsidy per gallon} = 1 - 1.5 \times I/P$$

where I is cost of inputs, and P is price of ethanol.

The subsidy is the highest when cost of inputs is low or price of ethanol is high or vice versa. The 1.5 only normalizes the long-run average of I ($2/gallon) to P ($3/gallon). Figure 11.28 shows the result of valuation when such a subsidy is applied. The value of the plant after construction increases to $101 million, and the value of the investment opportunity at start is a handsome $7 million and is sufficient to induce investment. The cost of this type of subsidy is substantially lower than the original blending subsidy. In this case, the plant does not operate when the price of ethanol is lower than the cost of production, and there are no production distortions due to subsidies. By inducing production when input prices are low or ethanol prices are high, the subsidy is applied only when production adds value. The plant operates only 77% of the time when ethanol production is profitable with or without a subsidy.

Another type of subsidy the government could consider is providing an incentive to increase efficiency in conversion through innovation. Consider a scenario in which the government provides a bonus:

$$\text{Bonus per gallon} = \text{price of ethanol}/(\text{cost of conversion} \times 100)$$

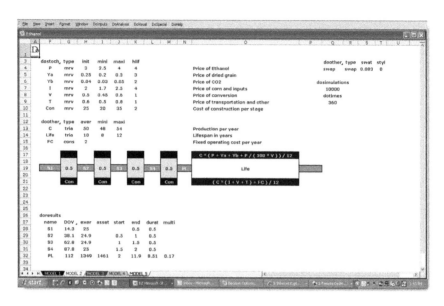

FIGURE 11.29
Model of the ethanol plant with subsidies based on conversion efficiency.

The lower the cost of conversion, the higher the bonus from the government. If the ethanol producer invests into process innovation and finds better ways to manufacture ethanol, the producer gets two benefits: the cost of production goes down and profit goes up and profits increase even further, enhanced by the bonus from the government. The result from such a bonus scheme is indicated in Figure 11.29. The result shows a healthy economic value of $14 million, sufficient to induce investment. Once the plant is operational, it will have ongoing incentives for the producer to drive down conversion costs by better design of equipment, process, and other technologies.

Another way to induce societally beneficial behavior by the ethanol producer is providing incentives for increasing the yield from the manufacturing process. Another bonus could be designed to induce investments into process technology that will enhance the yield from inputs (thus reducing the amount of input needed per gallon of ethanol produced). Although such incentives exist for the manufacturer naturally (as this will reduce the manufacturer's costs and increase profits), existence of blind subsidies such as a blending subsidy for every gallon produced makes it less likely that efforts will be expended in improving efficiency. Ethanol producers may spend more time creating bigger and less efficient plants as fast as they can as the profits from a live plant are so much greater (due to subsidies) than incremental benefits gained from efficiency improvements in existing plants. As the policy makers toast their success of subsidies, more inefficient ethanol plants will be constructed, each competing for inputs and driving input prices up. Since the inputs are also feedstocks, higher ethanol production increases the price of food for people as well as animals. The increase in cost of food for animals in turn drives up the cost of meat. Further, the need for transportation of feedstock to plant locations and ethanol to the terminal may drive up demand for transportation and result in correspondingly higher transportation costs. As the costs of inputs and transportation increase, many plants will start to lose money (even after the generous subsidy). Politicians may reconvene to "solve the problem," and they may consider creative solutions, such as a higher blending subsidy.

Valuation of a Flexible Electricity Generation Plant

Electricity is a special commodity. It is needed for virtually every aspect of life in the modern world, but it cannot be stored effectively and at scale. Electricity has to be produced when it is needed. If the supply of electricity is not commensurate with demand, the price will spike as the marginal user will pay the market-clearing price. Since total demand cannot be precisely predicted most of the time, demand and supply imbalances and corresponding price fluctuations do occur. Electricity producers deal with it with a network of plants, some producing power constantly (base load plants) and some operating only when there is a spike in demand (peaking plants). By optimizing production between the base load and peaking plants, generating companies can closely match the demand.

Base plants typically cost less and they operate for long periods of time before a planned or unplanned outage occurs. Nuclear power plants fall into this category. Peaking plants are flexible and can be switched on and off as needed. They are typically fossil power plants that use fuels such as natural gas or oil.

Valley Municipal Electric (VME) is considering investing into a new electricity generation plant. Its service area has been growing, with many manufacturing companies moving in and taking advantage of low-cost land and lower taxes. VME already has a few base load plants to meet the demand of its customers. It has been managing peak loads by buying electricity from others and enters into forward contracts for such purchases. As demand grows and volatility increases (due to many different types of manufacturers), VME finds it difficult to manage the peak loads. More important, the chief executive officer (CEO) felt that VME could take advantage of the situation and increase profits by adding a peaking plant to the network:

Plant type	Combined cycle gas-fired turbine plant with natural gas
Capacity	3,000 MW
Heat rate	5
Lifetime	20 years
Investment	$100 million

Heat rate is the conversion efficiency of the plant. The lower the heat rate, the more efficient the plant is in converting heat from gas into electricity.

$$Electricity(MWHr) \times HeatRate = Gas(MMBtu)$$

The CEO is not fully convinced of the advantages of the peaking plant as he felt that base plants with contracts to meet the peaks are the best way to manage the system. However, he has been listening to some of the enterprising engineers and financial analysts who have been pushing investment in a flexible peaking plant. The investment needed ($100 million) is substantial, and it required convincing both the CEO and the investors.

An electrical engineer decided to pull together an economic valuation of the proposed plant. In building a model, the engineer first collected historical prices for electricity and gas. In the case of electricity, grid congestion and associated spikes have been a common occurrence in the area. From historical prices, the engineer estimated the following factors for the electricity price process:

Long run mean	$60/MWh
Daily volatility	15%
Half-life	Half day
Probability of a spike (up or down)	10% per day
Magnitude of spike	20%
Standard deviation of spike	25%

The factors are represented in the form shown in Figure 11.30 in modeling the mean reverting price process for electricity.

Then, electricity prices for 20 years were simulated with a time step of a day. Figure 11.31 shows the result from the simulation. The engineer found that prices can exceed $120/MWh for many days during the course of the year. Prices also fall into the $40/MWh range routinely. Also, a forward curve exists for electricity, and this information is not currently incorporated into the simulation. Based on the forward prices, a forward curve for electricity is

dostoch	type	init	vola	mean	hlif	jpro	jmag	jsdv
electric	mrl	60	15	60	0.5	10	20	25

FIGURE 11.30
Stochastic characteristics of electricity.

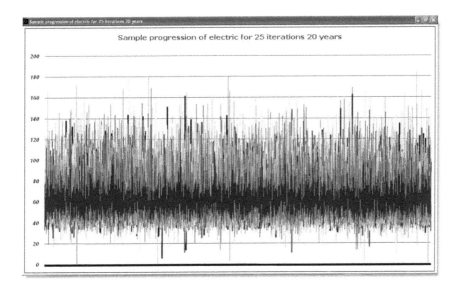

FIGURE 11.31
Stochastic simulation of electricity prices.

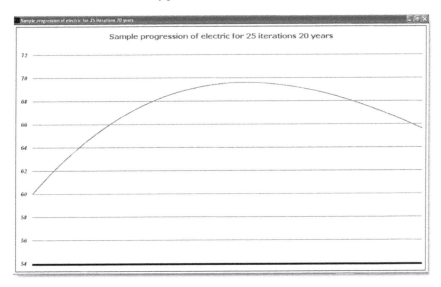

FIGURE 11.32
Forward curve for electricity.

shown in Figure 11.32. The forward curve indicates that the market expects the price of electricity to rise in the next decade with a subsequent fall to a more moderate level. Incorporating the forward curve information, electricity prices were simulated again, obtaining the chart shown in Figure 11.33. Although incorporation of the forward curve information can be seen in the

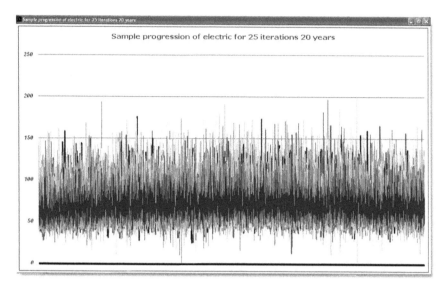

FIGURE 11.33
Stochastic simulation of electricity prices with forward curve.

simulation, clearly it is lost in the high daily volatility and spikes in prices. Nevertheless, the engineer was now satisfied that all available information was incorporated in the price process.

The next task was to model the gas prices. Being a commodity, gas also shows mean reversion. Based on historical data, the following parameters were estimated for the price process for gas:

Long-run mean price	$10/MMBtu
Daily volatility	5%
Half-life	10 days

Using simulation, the sample in Figure 11.34 for the price process of gas was obtained.

However, this was not enough. The engineer felt it important also to incorporate the forward curve information in the case of gas. Gas prices have been on the rise, and the market expects this trend to continue. Using market data, the forward curve for gas was constructed as shown in Figure 11.35 and could be incorporated into the simulation as well. Figure 11.36 shows the result.

The power plant could now be modeled. First, the plant was modeled as a base load plant in which the operation of the plant will be continuous. This means that regardless of the prices or electricity and gas, the plant will operate. The plant will lose money if revenue from electricity is less than the cost of generating it. Otherwise, it will make positive profits.

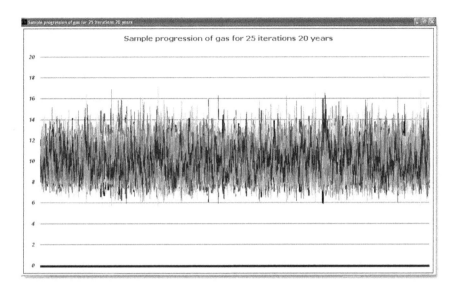

FIGURE 11.34
Stochastic simulation of gas prices.

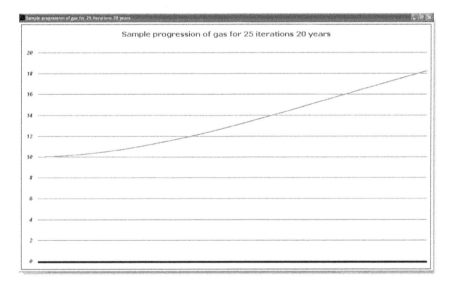

FIGURE 11.35
Forward curve of gas.

Using a real risk-free rate of zero, the risk-neutral payoff (Figure 11.37) that shows the value of the plant to be $51 million was obtained. This means that if the plant is operated as a base load plant with continuous operation, the expected profit from the plant for 20 years is $51 million. The plant is expected to make anywhere between $13 and $89 million during its lifetime.

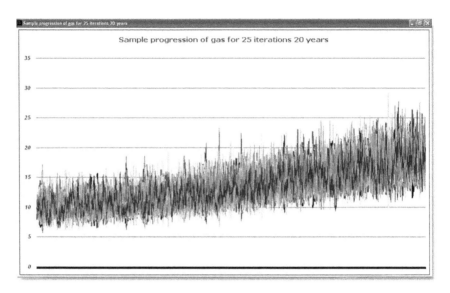

FIGURE 11.36
Stochastic simulation of gas prices with forward curve.

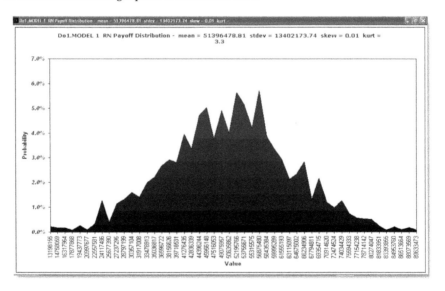

FIGURE 11.37
Risk neutral payoff from a continuously operating base plant.

Since the expected gain from the plant ($51 million) is less than the initial investment of $100 million, this plant is too expensive to run as a base load operation.

However, if the plant is designed and operated in a flexible fashion that allows the plant to be switched on or off every day depending on the price

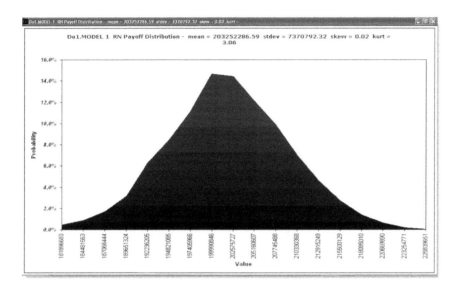

FIGURE 11.38
Risk neutral payoff from a flexible plant.

levels of electricity and gas, it may become viable. Every day, the plant manager will make a decision whether to operate the plant. If electricity prices are higher than heat rate times gas prices, the manager will operate the plant. If not, the plant will be shut down until the next day, when another decision will be made. Assuming the starting, shutting down, and maintenance costs are small, the engineer obtained the following result: the risk-neutral payoff shows the value of the plant to be approximately $200 million (Figure 11.38). The risk-neutral payoff shows a range of $180 to $250 million.

Figure 11.39 shows the model of the plant and the results. The plant will be operational only 11.3 years out of the available 20 years. This means that the plant will operate only about 200 days per year (on average) when the revenue from the plant is expected to be positive. As the plant nears its life end, the gas prices are expected to continue to rise (as indicated by the forward curve), and electricity prices are expected to fall. Thus the plant is likely to be operated less toward the end of its life.

After considering the results, the engineer wanted to make them more realistic. He knew that there are two types of outages that could happen in the plant: planned and unplanned. The plant was estimated to be in planned outage for 1% of the time for maintenance. This represents approximately 1 day per quarter. In addition, an unplanned outage (the plant shuts down for unexpected reasons) was estimated for 1 day per month. This represents a 3% probability of an unexpected plant shutdown on any given day. In combination, the planned and unplanned outages represent a 4% probability that the plant will not be functional on any given day. To represent the outage, a stochastic variable was created that assumes a value of 1.0 (normal

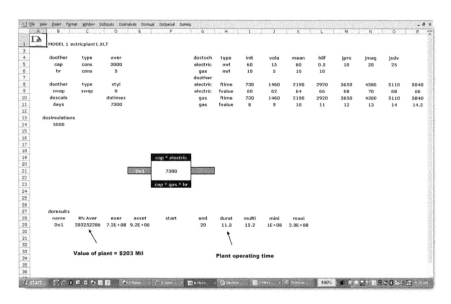

FIGURE 11.39
Model of the flexible plant.

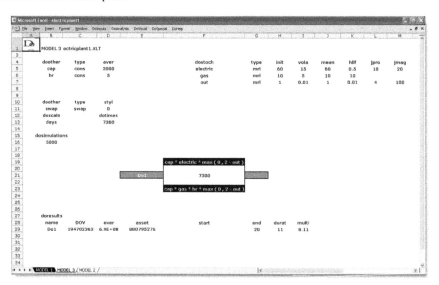

FIGURE 11.40
Plant model with outages.

plant operation) for 96% of the time and a value of 2.0 (plant shutdown) for the remaining 4% of the time. The model was run again (results shown in Figure 11.40), obtaining a value for the plant of approximately $195 million. So, the outages represent a cost of $8 million (difference from the original

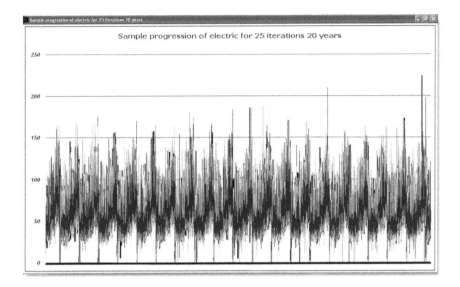

FIGURE 11.41
Simulation of electricity prices with seasonality.

value) for the plant owner. This is roughly the same as a 4% drop from the original value of $203 million.

The engineer wanted to add more reality to the model. For example, he noticed seasonality in the price process for electricity. By analyzing past prices, he found that the price in the summer was 25% higher than that in the spring, and price in the winter was 25% lower than in the spring. The fall price was approximately the same as that in spring. This seasonality in electricity prices was incorporated by modifying the long-run average (which is currently driven by the forward price curve, which is only available for yearly intervals). Figure 11.41 is a simulation of electricity prices when seasonality is included. The analyst found that the value drops to approximately $160 million, so seasonality in electricity prices is an important consideration in the value of the plant.

The engineer wanted to make a few more refinements to the model. First, the analyst wanted to consider volatility of electricity and gas prices as stochastic (time varying) rather than constant. The reason is that volatility regimes persist, and this may show serial autocorrelation. Generally, high volatility will be followed by high volatility until the regime breaks down, switching to a low-volatility regime.

He modeled the volatility as a mean reverting process with a half-life of 30 days. This means that it takes about 30 days for volatility to retrace its path back to long-run average after a significant excursion from the average. Figure 11.42 shows the simulated volatility in electric prices over the lifetime of the plant.

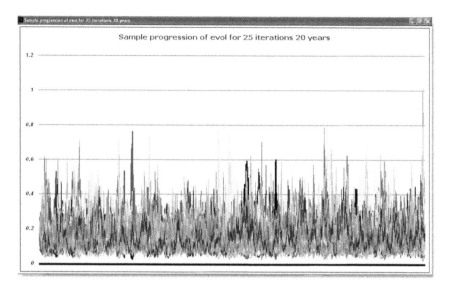

FIGURE 11.42
Stochastic simulation of electricity price volatility.

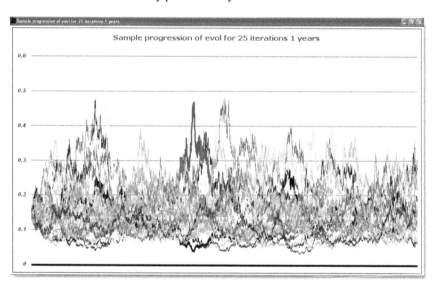

FIGURE 11.43
Volatility of electricity prices for a sample year.

Figure 11.43 indicates the progression of volatility during the course of a sample year. Slow mean reversion in volatility of 30 days implies that the higher or lower volatility tends to persist.

Figure 11.44 presents the result of the valuation of the power plant with all assumptions included. The present value of the peaking plant of $165 million

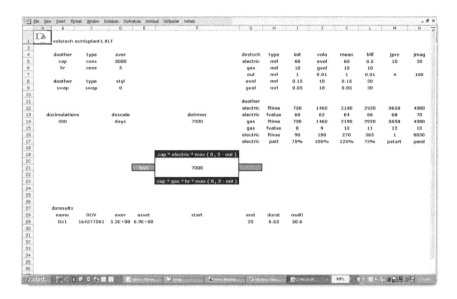

FIGURE 11.44
Model for the plant with all assumptions included.

exceeds the present value of construction cost of $100 million. Engaging in the plant investment will net the company $65 million in shareholder value. The analysis was good enough to convince the CEO of the merits of the project, and the company decided to proceed with it.

Although none of the assumptions—prices, seasonality, volatility, and other aspects—is known precisely, describing them with known uncertainties is an important aspect of the valuation of the plant. How the plant operates also significantly affects value. Existence of uncertainties is not a sufficient reason to avoid a systematic valuation of the plant as it is possible to value uncertainties as demonstrated in this case.

Opening and Closing of a Mine

The operator of a mine has the option to operate or temporarily close a mine, depending on the cost of the mined commodity. This can also be thought of as the flexibility to switch between operating modes—open or closed. Depending on the state of the mine, the switching of the operating modes may incur expenses. For example, opening of a closed mine may incur setup costs, and closing of an operating mine may incur closing costs. Neither opening nor closing costs will be present if the mine were operating continuously. Opening costs may be related to the removal of maintenance

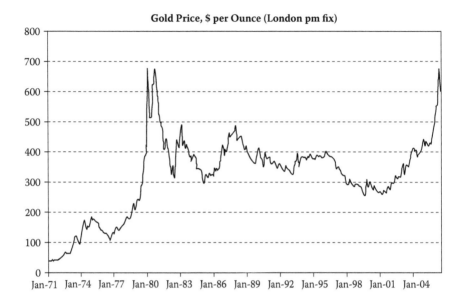

FIGURE 11.45
Gold prices.

equipment, addition of mining equipment and personnel, and other administrative costs. Closure costs may represent preparations to keep the mine idle, employment contract termination costs, and other regulatory and government costs. A closed mine also will involve additional recurring costs, such as maintenance because closing does not mean abandonment. In the case of closure, the mine has to be kept in a condition that will allow it to be opened at a future time. There may be regulatory or lease-related costs that may prescribe certain payments for environmental or local employment reasons. Since these costs exist, the decision to close, open, or maintain a mine in any given year is related not only to commodity prices and costs of production but also to future expected states of the mine. The cost incurred in any year is also affected by the current state of the mine (whether in a closed, open, or mothballed state).

Consider a gold mine the operator will only operate if the price of gold is greater than the cost of extraction, refining, and logistics. The operator will continue operating the mine in this case if it is already operating. If the mine is in a closed state, the operator will open the mine and incur an opening cost. If the cost is more than the revenue that can be generated from the produced gold, the operator will close the mine if it is operating or keep it closed if it is not operating. If the operator closes the mine, closing costs will be incurred. During the time the mine remains closed, it will also incur maintenance costs. Closing is a complex decision that must take into account the current state of the mine and additional costs incurred, if any, in closing the mine now and opening the mine, if needed, at a later time. Gold is extracted at a prescribed rate. This extraction rate in combination with the expected stock of gold in the mine will define the total number of years the mine can operate.

Figure 11.45 shows the non-inflation-adjusted price of gold from 1970 to 2005. As is obvious, gold prices accelerate rapidly when inflation surfaces as gold is considered an inflation hedge. High inflation in the 1980s is revisited in 2008, with growth accelerating in the developing countries and a depreciating U.S. dollar due to a variety of macro- and microevents.

The annual volatility of gold prices over the past 35 years is shown in Figure 11.46. Volatility takes brief but pronounced excursions from an average of 13% per year. Consider a mine with the following parameters:

Current price of gold	$900/ounce
Long-run mean price of gold	$500/ounce
Cost of extraction/refinement/logistics	$500/ounce
Annual volatility of price	13%
Annual volatility of cost	5%
Real risk-free rate	0%
Half-life of gold prices	1 year
Capacity	100K ounces/year
Total reserves	1 million ounces
Total time the mine is leased	20 years
Cost of opening a closed mine	$1 million
Cost of closing an open mine	$2 million
Cost of maintaining a closed mine	$0.5 million

Two mean reverting stochastic variables are used, one for revenue and the other for cost. Each year the mine is in operation, the revenue from the extracted gold is the price of gold times 100K ounces. At the current price of $900/ounce, this represents revenue of $90 million/year. However, the long-run average price of gold is $500/ounce, so in the long run, this revenue is expected to decline to a more moderate $50 million/year. The total cost of production in an operating year (which will produce 100K ounces) is also $50 million year (at $500/ounce). The total cost of production is equal to the expected long-run revenue from the mine.

Figure 11.47 is the model of the mine that allows switching flexibility (open, close, or maintain) in any year for a period of 20 years. A constraint is also applied to restrict the total number of operating years to 10 as the maximum stock of gold in the mine is only 1 million ounces and the rate of extraction is 100K ounces/year. Revenue from the mine in any operating year is modeled as G, and the total operating costs if the mine is open is modeled as E. The current revenue from the mine is $90 million, helped by the high current price of gold of $900/ounce. This revenue shows a volatility of 13% (same as the volatility of gold prices) and is mean reverting with a half-life of 1 year.

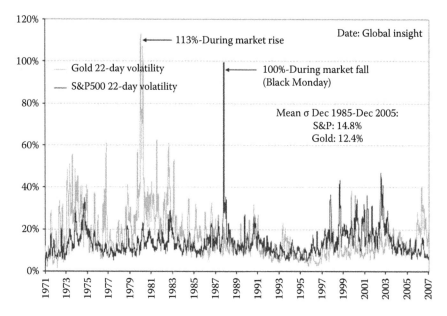

FIGURE 11.46
Gold volatility.

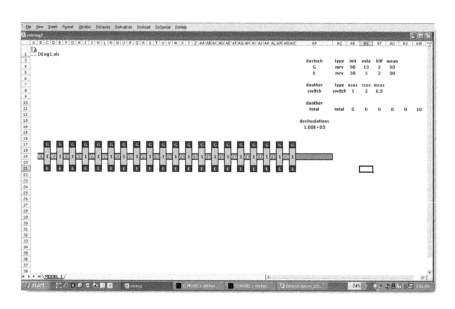

FIGURE 11.47
Mine model.

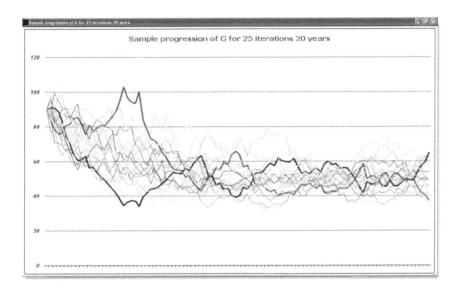

FIGURE 11.48
Stochastic simulation of yearly revenue from the mine.

The cost of production per year for the mine (if it is in the open state) is $40 million. The cost also shows a volatility of 5% driven by labor, equipment rental rates, cost of chemicals, and other costs.

Figure 11.48 and Figure 11.49, respectively, present the sample simulations of revenue and the cost of production from the mine over a period of 20 years. The costs of opening, closing, and maintaining the mine are also represented in the model. Every year, the operator of the mine makes a decision whether to operate the mine. The decision depends on the current and anticipated revenue from the mine, operating cost, and the current state of the mine. The current state is important in the operating decision because of the closing, opening, and maintenance costs. If the mine is open, the operator has to consider the continuation of the operating mine or incur the cost of shutting it down as well as the future maintenance costs. In certain cases, the operator may decide to continue operations even if the mine is losing money currently. This is because the anticipated future profits may be higher from the continuously operating mine than the profits from the one that is shut down and opened later. Similarly, a shutdown mine will remain inactive even if the current profits of an operating mine are positive. This is because the anticipated future losses from the open mine may be higher than the losses from a shutdown mine that incurs maintenance costs.

Assuming that the operator of the mine makes the right decisions every year, after observing gold prices and costs, we can value the mine. The risk-neutral payoff from the mine is shown in Figure 11.50. The mine is valued at $110 million.

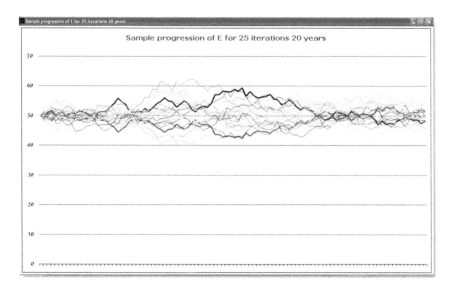

FIGURE 11.49
Stochastic simulation of yearly cost of operating the mine.

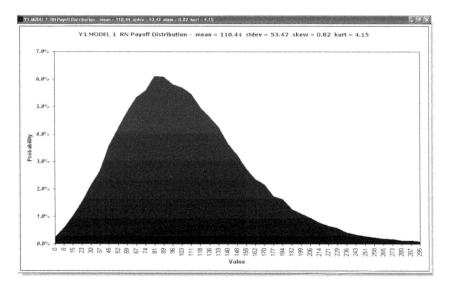

FIGURE 11.50
Risk neutral payoff from the mine.

As seen in the historic patterns of volatility of gold prices, volatility shows both regime persistence and extreme jumps for very short periods of time. Such jumps occur in periods of extreme tension or confusion in broad markets as investors flee to the perceived safety of real assets such as gold that serve as an inflation hedge. Volatility can be modeled as a stochastic variable.

dostoch	type	init	vola	hlif	mean	jpro	jmag	jpos	jsdv
VV	mrv	0.13	10	0.08	0.13	5	100	100	0.05
G	mrv	90	VV	2	50				

FIGURE 11.51
Modeling assumptions for revenue and revenue volatility.

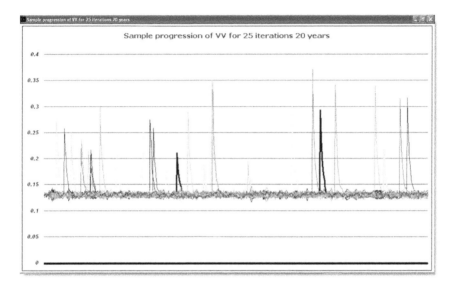

FIGURE 11.52
Stochastic simulation of gold price volatility.

Figure 11.51 indicates the parameters for a more detailed modeling of revenue from the mine with the volatility of revenue modeled to be stochastic. A variable VV is modeled to have an initial and long-run average of 0.13. The volatility of this parameter is 5%, and it is mean reverting with a half-life of 1 month (0.08 years). It also shows a 10% probability of a jump (an extreme and sudden move from the current value). When a jump happens, we expect the magnitude of the jump to be on average 0.13 (the total value doubles, modeled as a 100% jump) with a standard deviation of 0.05. Sample simulations of the volatility VV of the revenue from the mine are shown in Figure 11.52. With the stochastic volatility modeled in, the value of the mine increases to $112 mil (approximately a 2% increase from the initial $110 million).

In the valuation of the mine, thus far, we have assumed that the revenue from the mine and the cost of production are uncorrelated. Labor and equipment costs specialized for mining tend to increase when gold prices increase. It may be more realistic to assume that they are correlated rather than perfectly independent. Figure 11.53 presents assumptions with a 50% correlation assumption between revenue and cost. The first block defines parameter VV as stochastic volatility of revenue (which depends on gold prices). The

dostoch	type	init	vola	hlif	mean	jpro	jmag	jpos	jsdv	csto	corr
VV	mrv	0.13	10	0.08	0.13	5	100	100	0.05		
G	mrv	90	VV	2	50						
E	mrv	50	5	2	50	0	0	0	0	G	50

doother	type	ocos	ccos	mcos
switch	switch	1	2	0.5

doother total	total	0	0	0	0	0	10

dosimulations	dotimes
1.00E+04	1.00E+03

FIGURE 11.53
Mine model assumptions.

revenue parameter G takes VV as its volatility. Finally, cost E is described as correlated with G at the 50% level. Intuition tells us that when options are present (in these cases, they are switching options), any correlation between asset and cost will reduce value. It is the case for the mine value also. If costs move in correlation with revenue (and gold prices), the net profit from the mine when it is operational will be lower as higher revenue will be met with higher cost. Note that it does not matter how much higher the costs are if the mine is not operating. Since the mine is likely to be operating when revenues are high, if costs are lower or at the very least noncorrelated with revenues, profits are likely higher.

Figure 11.54 shows the risk-neutral payoff from the mine with these assumptions. The value of the mine is reduced to $105 million.

Sometimes, the country that has allowed a commercial enterprise mine imposes constraints on the company. Examples of such constraints include minimum level of local employment, maximum profits, and maximum revenue from the mine. Consider a revenue constraint that imposes maximum revenue from the mine of $500 million. This means that once the company has cumulatively obtained total revenue of $500 million, the leasehold terminates, and the mine reverts back to the government.

Figure 11.55 gives the cumulative risk-neutral payoff from the mine with a revenue constraint of cumulative maximum revenue of $500 million. The value of the mine drops to approximately $92 million. Cumulative probability also shows that we are 85% confident that the mine is worth at least $50 million.

Let us also look at the value of the mine if the gold price currently is at the long-run average of $500/ounce. Note that in the previous valuation we assumed the current price to be much higher than the long-run average at $900 million. With a half-life of 2 years, this price is expected to rapidly move down, as shown in the previous sample simulations. Figure 11.56 presents the cumulative risk-neutral payoff from the mine if the gold price currently

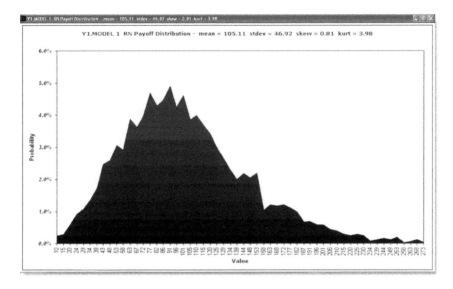

FIGURE 11.54
Risk-neutral payoff from the mine with correlation between revenue and cost.

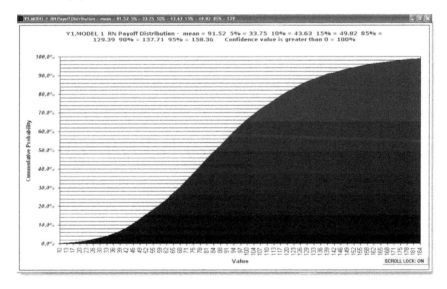

FIGURE 11.55
Cumulative probability of risk-neutral payoff from the mine with a hard constraint on total revenue of $500 million.

is at the long-run average price of $500/ ounce. The value of the mine is only $30 million in this case. This means that almost two-thirds of the $92 million value (calculated previously) is due to the current jump in gold prices. Under normal conditions, the mine will be valued much lower. An 80% jump in

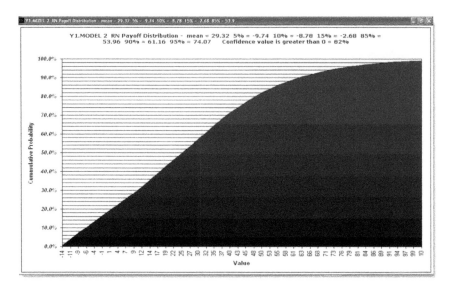

FIGURE 11.56

Cumulative probability of risk-neutral payoff from the mine with current gold prices at $500/ounce

gold prices (from $500/ounce to $900/ounce) has increased the value of the mine by over 300% (from $30 million to $92 million). Since the value of the mine is a derivative of the gold price (underlying asset), this is to be expected. Thinking back to the basics of option valuation, remember that the volatility of the derivative (option) is much more magnified from the volatility of the underlying factors.

12

Case Studies in Financial Services

Finance has been a hotbed for innovation for the past decade. However, most of the innovation has been limited to securitization and bundling of financial claims in a variety of ways. The forces that drove this innovation wave may also have taken us away from market-based economics, in which risk is always priced in a normative fashion. One thing we learned from the recent meltdown of the subprime markets is that there will always be participants who get lazy in assessing the risk and pricing the instruments they are buying and selling. Instruments that bundle risk are complex and if such instruments are transacted among a few players in somewhat inefficient markets, inefficient pricing can prolong for a period of time. However, if the risk is not priced properly, one party in the transaction will pay for it (if not now, then later).

Market-based pricing of risk in securities is as important in the public markets as in the private markets. The methodologies and tools discussed thus far are valid in both markets. Financial instruments and investment decisions, whether in public or private markets, share common characteristics in terms of the stochastic characteristics of underlying assets, private risks, and interacting components. It is important to systematically disaggregate the instrument or decision into components and characterize them appropriately as options, risks, and predetermined outcomes. Then, each of these components has to be analyzed using a consistent and systematic methodology that is market based.

Those who blame innovation in derivatives and hedging instruments for the creation and bursting of the dot-com and real estate bubbles as well as market volatility miss the following important notions:

1. Bubbles form in discontinuities when market-based valuation becomes difficult to perform using traditional techniques with models that do not capture uncertainty and flexibility. This is a good example of what happened in the dot-com bubble. A systematic market-based valuation by capital providers and investors may have avoided that problem. When market-based valuation is absent, buyers and sellers tend to use proxy pricing. The inefficient pricing of initial public offerings (IPOs) seen during the dot-com bubble is a good example of this. Similarly, the real estate bubble suffered from lack of consistent and appropriate valuation methods. So, a common symptom of bubble formation is that at least some participants in

the market are transacting without pricing the instruments using market-based techniques. One should also note that bubbles (as we define them) are not inconsistent with efficient markets. When uncertainty is high, as in a technology or asset price discontinuity, the expectations of market growth have a significant impact on prices. Even when the market incorporates these assumptions into prices properly, they may turn out to be incorrect.

2. Financial innovation is not a synonym for fraud. If market participants do not play by the rules, all sorts of problems can ensue. Regulators and compliance officers cannot fail to fulfill their responsibilities and expect well-functioning financial markets.

3. Derivatives and hedging instruments are fundamentally important for the functioning of financial markets. Hedge funds, for example, provide an important service to the markets by providing and removing liquidity as new information arrives. Volatility is a characteristic of the markets, and it will rise in periods of higher uncertainty even if the regulators lock up all hedge fund managers and speculators (assuming they can be identified uniquely from hedgers). Hedgers and speculators are necessary participants for well-functioning markets. Unfortunately, they do not wear color-coded jackets so that we can pick out the speculators from the crowds or open cry pits.

4. Leverage adds significant risk to an investment. Certain large firms were given exceptions by the regulators to increase leverage as much as three times normal (1:40), and they obliged instantly by buying large amounts of inefficiently priced security bundles using borrowed money. If the returns are not positive in the underlying bundle, then leverage will create significant distress to those on the bad side of the transaction. As one senior executive from an investment bank famously observed, "People are learning there are two sides to leverage."

In sum, the application of basic principles—market-based valuation of security bundles, consistent implementation of well-crafted regulation when market failures are present, and holistic risk management with an understanding of the effects of leverage—would have avoided the financial crisis of 2008.

Option to Delay Decisions

Next, we evaluate decision options to delay decisions. Imagine a contract in which we have the option to swap a notional amount of $1 billion for €1 billion. The exchange rate currently is $1 = €1. The swap can be exercised at the end of each of the next 5 years. Each year, the buyer of this instrument has

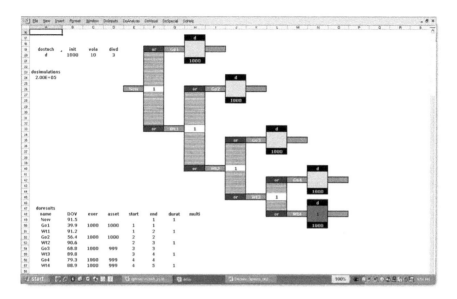

FIGURE 12.1
Currency swap model with option to wait.

a choice (1) to exercise the swap or (b) to decide to wait until next year. The buyer will only exercise if the value of the dollar is higher than the euro. Only then does the buyer have to decide whether to exercise or delay the exercise. If the value of the dollar is lower than the euro, the buyer will automatically wait until the next year and see if the dollar appreciates. The volatility of the dollar/euro exchange rate is 10% per year. The real risk-free rate is assumed to be zero.

The model and result are given in Figure 12.1. The exchange rate d (price of dollar in euros) is modeled using the geometric Brownian motion (GBM) process. We value this instrument at $91.5 currently. As you would have guessed, there is no value in exercising this option before the end of the full 5 years as premature exercise of an option is always suboptimal. The holder of this instrument will swap (if it is in the money) only at the end of 5 years (as indicated by the shaded swap). In effect, the decision to exercise will be delayed until the fifth year. When the holder of the swap feels that the exchange rate may move against him or her, the holder can simply sell the swap (assuming there is a market) and not exercise it.

Now, consider when the dollar is expected to depreciate at 3% per year. This can be modeled as a dividend since depreciation of the currency is similar to the stock giving out dividends (and declining in value). In this case, an earlier exercise may become optimal. The model and result are given in Figure 12.2. We value the instrument at $42 million with the dollar expected to depreciate at 3% per year. The expected exercise in this case is in year 4, as indicated by the shaded option.

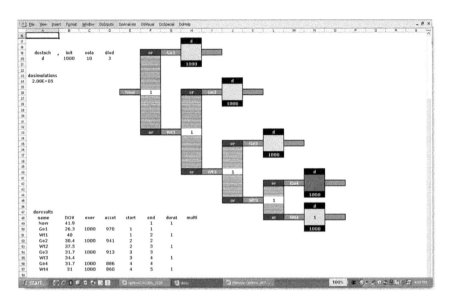

FIGURE 12.2
Currency swap model with slowly depreciating currency.

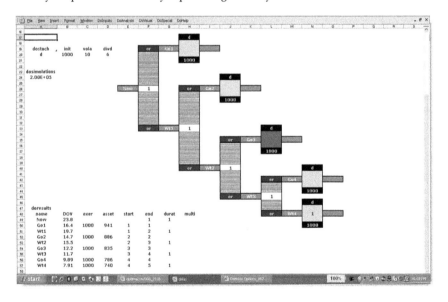

FIGURE 12.3
Currency swap with moderately depreciating currency.

Next consider an expected depreciation of 6% per year for the dollar; the model and result are given in Figure 12.3. With a higher dividend, the option to delay becomes less valuable. The instrument is valued at $24 million with an expected exercise in year 3.

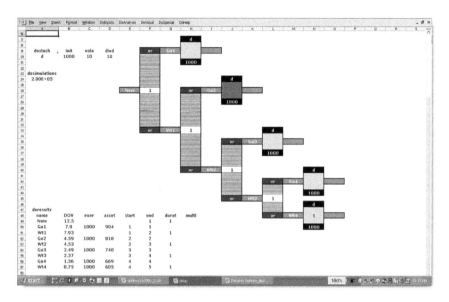

FIGURE 12.4
Currency swap with heavily depreciating currency.

Now, consider an even higher depreciation of 10% per year. The model and result are given in Figure 12.4. As expected, the value drops dramatically to $12.5 million, and now the expected exercise is in year 2. The value from the option to delay exercise is even lower.

This example demonstrates the option value to wait and see how the value of the delay option declines in the presence of a dividend or depreciation of the asset. Depreciation of the asset can also happen in pharmaceuticals, for which the patent life is limited, and in mineral exploration and development, for which the life of a leasehold is limited. In pharmaceuticals, when patent life is limited, delaying an experiment means that the value of the marketed drug will be lower (because of the lower expected patent life when it enters the market), and this is similar to a dividend-paying stock. Similarly in minerals exploration and development, delaying exploration or development means that the length of time available for extraction is shorter; thus, the value of the developed mine is also less. This is also similar to a depreciating or dividend-paying asset.

Consider a pharmaceutical research and development (R&D) program in which a phase can be delayed as much as 5 years (Figure 12.5). Every year the program is delayed, valuable patent life is lost, and the value of the marketed drug drops by 10%. Also, every year the program is delayed, costs rise as much as 3% due to higher production and testing costs. Assume the value of the drug at the end of the phase is $100 million, and the cost of completing the phase is also $100 million.

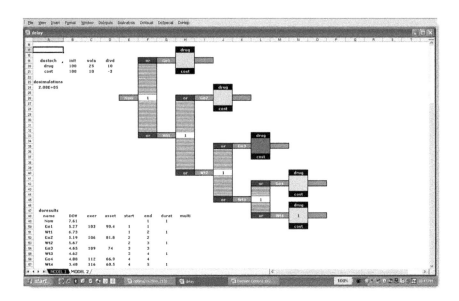

FIGURE 12.5
Delaying R&D program when timing flexibility is present.

The current value of the R&D program is approximately $8 million. We find that the program is expected to be delayed 3 years even though the marketed value of the drug is declining and the costs are increasing. Waiting to assess both the drug's marketed value and the cost has significant option value. Pharmaceutical companies in the past operated on the assumption that the faster they reached the market, the better. Although this may be true when marketing advantages exist for early entry, it may not be universally true for all programs. For example, assume that there is a 25% chance a competitor may enter any year in the 5-year decision horizon, and if an entry occurs, the value of the drug catastrophically drops by 50%. This provides additional incentive to start the R&D program early. The model and result are given in Figure 12.6. Value now drops to $5.5 million, and the program is expected to be started 1 year earlier than before (still a delay of 2 years in program start is optimal).

The exploration and development programs in natural resources are also similar. Figure 12.7 shows an example in which the developed mine is valued at $100 million. It is mean reverting with a half-life of 3 years. The cost of development is $50 million, growing at 3% per year. The investment opportunity is valued at approximately $52 million, and the expected start of development is in year 4.

This demonstrates that even when asset values are dropping continuously (due to fixed patent life as in the case of pharmaceutical R&D) and the cumulative probability of catastrophic loss increases with time (due to completive entry and other factors), there is significant value in option to delay decisions. Although this is generally well understood in exploration and development

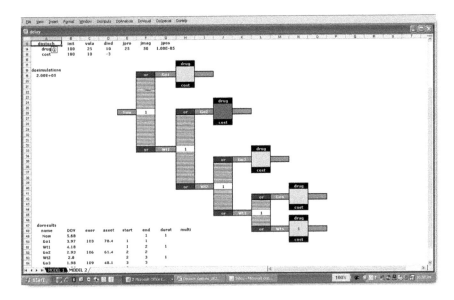

FIGURE 12.6
R&D start delay even in the presence of competitive entry.

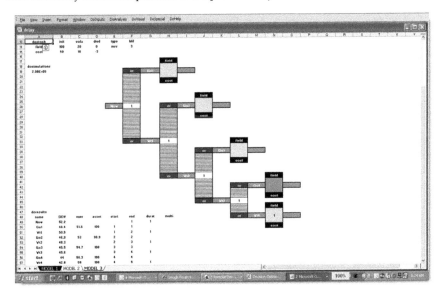

FIGURE 12.7
Delaying option in natural resources exploration and development.

of natural resources, it is much less of a consideration in life sciences, for which getting to market as fast as possible has been the normal mode of operation. This habit formed in the 1990s when the value of a marketed drug was very high (compared to costs), and the loss of value due to patent loss and competitive entry was significant. However, the economics of the industry have changed dramatically in recent times, and rushing to market at any cost may not be universally optimal.

Carve-Out Option for Business Unit

The technology company HTech has been considering spinning off one of its divisions to raise cash for future investments. This division, which manufactures mature electronic components, has operations around the world and an infrastructure that is well managed. The decision to spin off has been contentious among senior managers. Some felt that the division is well positioned for a spin-off. They made the following compelling arguments:

1. The new company can provide good transparency to Wall Street into its source and use of cash and thus allow it to raise financing if needed in the future.
2. The business is reasonably mature and, as a separate company, may be allowed to rev up its innovation engine and bring up new products to drive growth in the future.
3. The spin-off allows the parent company to raise cash quickly and redeploy it to other areas such as expansion of core products into emerging markets.

However, a smaller contingent felt that there are significant synergies between the components division and the rest of the company, including controls and robotics. They argued that many of the design engineers work across the divisional boundaries and share ideas. Further, a detailed understanding of component costs allows engineers to design more efficient controls and robots, the company's core products. After many months of debate, the management team was nowhere close to a final decision, and the investment bankers they engaged to advise were getting impatient.

One of the analysts in the corporate finance department had an interesting idea to possibly bring the two sides together. She suggested that the company consider a "carve out" of the division with an option to buy back control if the "synergies" that were argued to exist were significant. In effect, the company can spin off the division but hold an option to buy back 51% of the equity (to gain control) at a predetermined price within a period of time. This is a simple call option. To effect such a transaction, the company may have to pay the investors something close to the value of such an option. It may be able to do this by providing a discount to the initial price.

In a typical carve-out, only a minority stake in the new company is provided for the IPO, and the parent company holds the majority stake. In this "special carve-out" suggested by the analyst, all the equity is provided for the IPO, but the parent company holds an option to buy back a majority stake in the future. This option is valuable and may have an effect of capping the returns to the IPO investor.

The mechanics of the option exercise will be as follows. If the parent company exercises the buyback, the new company will issue new shares that will be bought by the parent company at a prespecified price. The parent company is likely to exercise this option only if the price of the stock of the new company is higher than the prespecified strike price, and this acts as a cap on the stock price of the new company. Hence, the exercise of this option may mean dilution to the existing shareholders of the new company. The chief financial officer (CFO) was intrigued by the idea; however, it was not clear to him how the company could value such an option. The company will only exercise the option if the value of the division including the synergies is greater than the strike price. He decided to compile estimates of the various parameters. Using simple price/earnings (P/E), price/sales (P/S) comparables, he valued the division at $100 million currently. The synergy between the components division and the rest of the company will only be evident over time after the division is carved out and has started operating as a separate company. He estimated this synergy as a multiplier on the value with expectation of 1.1 (10% enhancement in value of the division) but in the range of 1.0 to 1.25. A value of 1.0 means no synergy. This multiplier can only be captured by the parent company (as the synergies are between the different divisions in the company) and can only be observed by it.

After discussions with the corporate finance team, the CFO decided to set the strike price at $100 million and the option as European type with duration of 5 years. This means that the parent company has the option to buy 51% of the carved out division in 5 years for $100 million. It will not be a tender offer, but the new company will issue new shares to the parent company so that the parent company will own 51% of the outstanding equity and the new company will book $100 million in additional paid-in capital. So, postexercise, the new company's value will be enhanced by $100 million. The company has no debt in its capital structure. The CFO also decided to use a real risk-free rate of zero.

At IPO, the company is valued at $100 million. The analysts expect the value of the new company to be in the range of $50 to $200 million in 5 years. Figure 12.8 is the model that shows the call option. The "asset" that can be bought in 5 years is 51% of the new company, and the $100 million is paid in capital (since new shares will be issued) enhanced by the factor called synergy. The value of this option today is $15 million. Since this option is effectively "written" by the investors of the new company at IPO, the company has to pay the investors $15 million for the transaction to be value neutral. At an IPO price of $85 million (a 15% discount to the current value of the division of $100 million), the transaction will be value neutral. Since this information

FIGURE 12.8
Carve out with option to buy back.

is known only to the company and the synergies can only be captured by the company, the CFO felt that he could market the IPO without a discount. Cosmetically, since the strike price of $100 million is half of the high-end value expected for the new company in 5 years, he felt that the investors would value the call option lower than what it was actually worth.

If the company were to hold 51% of the equity at IPO, as is customary in a carve-out, the IPO will raise $49 million (for the 49% equity), which does not capture any incremental value for the company's shareholders who also stand the risk of losing value if significant synergies exist between the spun-off division and other parts of the company. Instead, in the nontraditional carve-out as suggested by the analyst, the company can enhance shareholder value by better marketing of the following features and issue an IPO without discount (for the call option):

1. The current value of the division is $100 million. In 5 years, the value of the division (as an independent publicly traded company) will be in the range of $50 to $200 million.

2. The company intends to spin off the division for $100 million with a special feature. It can buy 51% of the new company in 5 years for $100 million. This is the upper end of the range of expected value for the new company in 5 years, so the original investors do not lose anything.

3. When it buys the 51%, additional shares will be issued, and the company will invest an additional $100 million in the new company, helping its liquidity position tremendously.

4. In essence, this option is like a positive boost for the investors in the IPO. If the new company is successful, it is guaranteed a capital infusion by the parent at a price that is expected to be the highest value for the independent company. The parent not only is willing to spin off and sell the entire division but also is going to "stand behind" it with an option to provide a capital infusion in the future. The investors should gladly give this option to the company.

By convincing investors of this argument, the CFO enabled the current shareholders of the company to pocket $15 million in incremental value. It also allowed the company to keep its option free and open. In 5 years, the company will get a lot more information on the synergies as well as the economic prospect of the spun-off division. If everything is to its satisfaction, it can gladly invest the $100 million and take control of the spun-off division. If not, it can simply walk away. The IPO was oversubscribed, and the price rose 20% on the first day. The option feature "was deemed attractive" by certain institutional investors and hedge funds.

Economic Valuation of a Private Company

LiveRx is a small biotechnology company established by two university professors. Over the last couple of years, they have been perfecting a platform technology (CytoRx) that enables companies to create more potent and safer drugs. First, they planned to license the platform technology to other life sciences companies, which could incorporate it into their R&D processes to improve productivity. The productivity benefits come from lower attrition in the pipeline as CytoRx allows companies to improve the characteristics of their chemicals and improve the therapeutic index (higher efficacy and lower toxicity). As they were applying the final touches on version 2 of the technology, they began to think differently about their new company. They wanted to develop a marketing infrastructure and raise additional capital.

To seek capital, they had to produce a premoney valuation of the company. Since the company has no cash flows currently, its value resides in a patent (for the platform technology) and the passion of two individuals to create a new business. The LiveRx founders felt that the traditional net present value (NPV) analyses would not be that useful in assessing the value of their enterprise.

At a biosciences conference in Chicago, one of the founders met an old friend who agreed to help them value LiveRx. She said that a decision options analysis may be suitable as there is significant uncertainty and decision flexibility in new chemical entity (NCE) research and development. In consultation with the founders, she collected the following assumptions:

1. The LiveRx technology platform is applicable only in oncology (cancer treatment). The R&D program for cancer pursued by life sciences companies has the following major phases in sequence:

    a. Lead: A lead chemical is identified for a specific target.

    b. Preclinical: A candidate is nominated for preclinical experiments.

    c. IND: Investigative New Drug application is filed

    d. POC: Proof of concept is established

    e. Development: Full-fledged development to prove viability occurs.

    f. NDA: A New Drug Application is filed.

    g. Launch: The new drug is launched.

2. They expect each NCE to spend 9 to 18 months in each of the phases, except in the development phase, which is expected to be twice as long. Launch will be immediate after approval. The durations of these phases are correlated. The more time it takes to complete one phase, the more time it will take to complete subsequent phases.

3. Currently, the probability of success in each phase is 75%, except in NDA, which has a probability of 80% for approval. After approval, there are no private risks. With the LiveRx platform, companies can reduce failure rates in each of the R&D phases except in NDA, which is governed by the whims of the regulators. So, with the LiveRx platform, companies can expect a success rate of 80% in all phases. The success rate in NDA will remain the same at 80%.

4. They expect the following costs for each of the NCEs, with all costs correlated 100% except launch costs, and the more it costs in one phase, the more it will cost in subsequent phases:

    a. Lead: $0.75 to $1.25 million, average $1 million.

    b. Preclinical: $5 to $8 million, average $6.5 million.

    c. IND: $15 to $25 million, average $20 million.

    d. POC: $25 to $45 million, average $32.5 million.

    e. Development: $75 to $ 150 million, average $100 million.

    f. NDA: $20 to $40 million, average $26.5 million.

    g. Launch: $75 to $125 million, average $100 million.

5. Cancer indications, as targeted by LiveRx candidates, are expected to have the following characteristics in the market:

    a. Peak sales: $200 to $450 million, average $300 million.

    b. Net margin: 40% to 50%, average 45%.

    c. Time in the market: Approximately 10 years.

Based on these assumptions, she calculated the NPV of the drug when it enters the market to be $425 million with a range of $300 to $650 million. In doing so, she used the cost of capital derived from pure play companies in the area of cancer therapies.

6. The real risk-free rate is zero.

She first constructed a decision options model with these assumptions and status quo success rates (Figure 12.9). The results table shows the economic value of oncology drug candidates of this type in various phases. She ran the analysis again with the expected success rates with the application of LiveRx technology. The summary of the results is given in Figure 12.10.

In Figure 12.10, the column "Before" shows the value of the candidate before the application of LiveRx (in respective phases), and the column

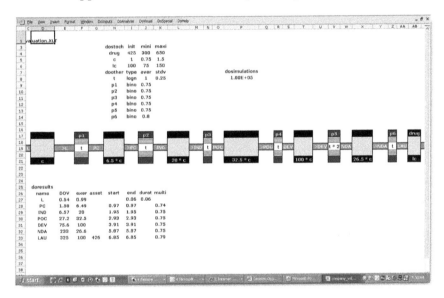

FIGURE 12.9
Oncology R&D model.

Phase	Before	After	Difference	% Gain
L	0.54	4.36	3.82	707%
PC	1.58	6.68	5.1	323%
IND	6.57	14.7	8.13	124%
POC	27.2	39.1	11.9	44%
DEV	75.6	87	11.4	15%
NDA	233	233	0	0%
LAU	325	325	0	0%

FIGURE 12.10
Value of R&D program in various phases.

"After" shows the value with the application of the technology. The value is higher with the application of the LiveRx technology because of the higher expectation in success rates. As the numbers indicate, the percentage increase in value is higher earlier in the pipeline because of the combined effect of higher success rates in various future phases.

The difference in value of drug candidates in various phases is substantial with the application of LiveRx technology. The "Difference" column indicates the incremental benefit gained by LiveRx (prospective) clients by the application of its technology. In licensing the technology, LiveRx could consider a structure in which the benefits are shared between the client and the company.

For example, for a biotechnology company with five leads (L) and two INDs, the application of the technology will increase the value of the company by $5 \times 3.82 + 2 \times 8.13 = \$35$. The value of the company can be increased by $35 million by the application of the technology on the stock of candidates in the pipeline. If the client is willing to share 25% of this incremental value, the license will be worth approximately $9 million.

The LiveRx licensing structure can be based on when the technology is applied by its client (in which phase of the candidate), and each candidate will create only a single payment. LiveRx must now approach venture capital (VC) firms and debate the number of clients they can develop and more importantly how many candidates in different phases are likely to implement the technology. LiveRx partners identified a number of biotechnology companies that could use the technology in the next 3 years (they expected this to be the life span of the technology). Based on the pipeline of these companies, they estimated the number of candidates in development in different phases. Figure 12.11 shows the total number of candidates at client companies that they estimated could benefit from the technology, license fee per candidate, and total expected fee. Based on this, the LiveRx founders estimated that the platform technology is worth $50 million premoney. They wanted to raise $10 million from a VC firm for a 20% stake in the company.

They knew that the negotiations with VCs would be tough, so they wanted to start at a higher valuation for the company and negotiate their way down. After discussions with a few of their friends who had raised capital in the

Phase	Candidates	Licence/Candidate	License
L	10	1	10
PC	8	1	10
IND	5	2	10
POC	4	3	12
DEV	2	3	6
NDA	0	0	0
LAU	0	0	0
	Total		48

FIGURE 12.11
Calculation of fees from clients.

past, they decided to start at a valuation of $75 million. In the negotiations, if the VC valuations fell below the $50 million mark, they would not take the deal and would look for alternatives. The knowledge of this reservation price ($50 million in this case) is powerful in negotiations.

Decision Options and Financial Hedging

Multinational companies face exchange rate exposures. There are multiple types of exposures, such as translational, transactional, and operating exposures. *Translational exposure* results from translating revenue and costs that occurred in another currency to the home country (or reporting) currency. It is a tactical accounting exposure. *Transaction exposure* comes from contractual terms expressed in one currency and the movements in the exchange rates by the time the contract is executed. This is generally short term in nature. *Operating exposure* emanates from the organization and operations of a multinational company that generates revenue and incurs costs in many countries as the result of the company's strategy. Operating exposure is a long-term issue for any multinational company. Transaction and operating exposures are related and may have to be considered together.

Financial hedges are typically used to ensure against short-term moves in exchange rates. Southwest Airlines' purchase of forward contracts on fuel is an example of managing operating exposure, but such contracts can also be short term, ensuring against anticipated transaction exposures. Since such contracts have a fixed quantity, changes in underlying business (as in the case of a rapidly falling demand for air travel) may expose the company to exchange rate risk in a direction opposite what was carried prior to entering a forward contract. The longer the timeline associated with the exposure, the more difficult it is to manage such exposure with more tactical financial hedges. Even the best thought out financial hedges can ultimately result in exposures in a direction not anticipated by the company. Such events can be extreme and may even result in the failure of the firm.

To effectively manage currency exposures, multinational companies have to design and manage an estate of financial hedges and decision options (strategic and operating strategies) in a fashion that maximizes the value of the firm. An example of a decision option may be Toyota maintaining flexible manufacturing operations in Japan and the United States, allowing it to switch production to take advantage of exchange rates in the future. It is not only the more obvious location flexibility that is of importance here, but also other important aspects such as equipment flexibility (ability to switch production lines to different automobiles such as cars and trucks), labor flexibility (ability to stop and start production by ramping employees down and up), equipment/labor switching flexibility (ability to change labor content in the manufacturing process by incorporating more or less automation), energy flexibility (ability to use a variety of energy sources), and others.

Flexibility may also be designed in other parts of the operations, such as engineering, marketing, and logistics. The engineering design of manufactured goods should be a focus area for companies as the design can enhance the company's ability to take advantage of various types of flexibilities that may exist in the supply chain. For example, modular design and switchable components could provide the company opportunities to disaggregate manufacturing on an as-needed basis and the ability to shift production to different locations. Postponing final customization is an example of enhancing logistics flexibility. In this case, the company delays the final customization of the end product until it is ready to sell the product wherever demand is available. The uncustomized product can be shifted from a low-demand area to a high-demand area, thus enhancing marketing flexibility as well. Country-based branding coupled with regional brands that target a wide range of price points is an example of marketing flexibility that allows the company to move up or down the price/brand spectrum on an as-needed basis.

If Toyota proactively takes actions to establish a production network with such flexibilities designed in, its ability to take advantage of tactical exchange rate fluctuations increases, and the need for pure financial hedges (which can be expensive) decreases. However, it may still need to engage in financial hedges at least for the lead time in the execution of such plans. Because exchange rates are mean reverting, there is a "convenience yield" to actually having a flexible operating strategy that is active rather than "on paper." The same is true for other types of flexibilities, such as in engineering, marketing, and logistics. The lead time in implementing flexibility may be reduced by a financial transaction such as buying a company or existing assets of another company. However, in such a case, the company may have to forfeit (or the very least share) the value generated by flexibility as the seller is fully aware of such value. Collaborations and partial equity stakes in other companies in locations where the firm anticipates exchange rate exposures are other ways to increase flexibility without high up-front investment. One could argue that such actions also provide the company an option to delay larger investments. However, this may also have a price as the collaborator may seek features in the contract that may deter the firm from taking advantage of favorable future exchange rate regimes by abandoning the collaboration and going on its own.

It is often the case that corporate finance focuses on the financial aspects of the company, including the tactical financial hedges, and other departments such as strategic planning worry about decision options. Such a management configuration will certainly suboptimize the value of the firm as each will attempt to maximize only one piece of the puzzle. To make matters more complicated, there can be significant interactions between financial hedges and decision options in the company, and systematic portfolio management of the entire estate of financial and decision options is the only way to maximize value.

Traditionally, economists have argued that a firm should "hedge" all risks over which it does not have control. This will allow the managers to focus on those aspects of the business that they can influence. However, such a thought process may make managers focus only on tactical financial hedging. It is possible that managing anticipated risks only in some part of this connected system while ignoring others may in fact do more harm than good. In some quarters, "holistic risk management" has become a fashionable term, although its meaning is unclear. Often, it means all financial risks, as in the case of hedge funds. Managing financial and real risks together in a portfolio considering the interactions among them is true holistic risk management for companies. To accomplish this, firms not only may have to reorganize (away from traditional silo management of financial and real risks) but also implement tools capable of analyzing all risks in the same framework. Such tools should also be able to incorporate interactions between risks. In such a process, firms may also identify risk components that reinforce each other and when certain thresholds are breached may exhibit characteristics that cannot be controlled. It is this "runaway train" phenomenon and not the lack of traditional risk management that subjects most companies to the risk of failure. Tactical and financial risk management may keep them afloat when risks are not high, but such a process may not be enough when a discontinuity arrives.

Clearly, the scope of the firm is important in its ability to take advantage of a systematic management of financial and decision options. The larger and the more spread out the firm is, the higher the value of such a process. The currency regime in which the country is operating is also a factor. If the home country's currency is pegged and the firm largely transacts in a location where the base currency exists, it may not have a great reason to think about exchange rate exposures. If the firm's products and services have more commodity-like characteristics, it is more likely that it can take advantage of flexible manufacturing and logistics operations as location specificity is low. Even those firms with specialized products can move down from specialization to commoditization by breaking their products down to components and increasing configurability and switchability. Such a thought process, seeking flexibility in every aspect of the firm, has now become a necessary condition for survival and success in a connected world of people, products, and skills.

Enticing Infrastructure Development in Developing Countries

Some developing countries, such as India and Brazil, are in need of significant infrastructure development and improvement. The government can entice commercial enterprises to partake in such activities through a combination of subsidies, guarantees, and other incentives. One such area is in build, operate, and transfer (BOT) schemes for highway construction projects. The government awards a BOT contract to a commercial enterprise, typically through a bidding process. Let us analyze such a project.

Consider a highway construction project connecting two cities. The length of the required four-lane road is approximately 200 km. MaxCon is a leading construction company considering bidding on the project. The project leader has already conducted surveys and prepared a detailed plan for the construction of the highway. She has estimated that the total cost of the project, which is expected to take about 4 years, is $75 to $100 million. The BOT contract package that is up for bid has a number of features:

1. At the end of construction (build phase), the company has the option to sell the highway back to the government for a fixed price of $90 million.

2. During the build phase, if MaxCon abandons the project, no repayment is possible, and the government will take over the project.

3. If the build phase takes more than 4 years, the company has to pay the government a penalty of $6 million for every additional year (or $0.5 million per month for the delay in completion).

4. During the operating phase (which is expected to last 10 years), the company can collect tolls for the traffic on the highway.

5. At the beginning of each year of the operating phase, the company has the right to demand its debt service cost (in lieu of tolls for that year) provided that the starting debt does not exceed $75 million and 10% of the principal is retired every year. If the company elects to do this, the government will collect the tolls for that year. The company can elect to receive this payment either in local currency or in the currency of contract transaction (dollars).

6. Every year during the operating phase, the government will guarantee tolls of $7 million. If the tolls are less than the $7 million, the government will pay the difference (top off revenue to at least $7 million/year).

7. During the transfer phase, the government will make a final payment to the company that is 10 times the tolls collected in the last year over and above the expected $10 million per year.

Using detailed traffic pattern analysis, the project leader has calculated that the tolls per year will be in the range of $5 to $15 million per year. The prevailing interest rate is 10%, and the company will capitalize the project with $75 million in debt and the rest in equity. To determine an effective bidding price, it must first value the project.

The contract attributes seemed like options to the project leader, so she asks the finance department to devise a decision options model. The following features exist in the model:

1. There is a "put option" to sell the entire project back to the government and walk away with $90 million at the end of the build phase. In this case, the company decides not to operate the project and the maximum loss it will incur in engaging in the project is the cost of construction, $90 million. The project leader's estimates showed that the project is going to take between $75 and $100 million for construction, so this is not necessarily a good outcome for MaxCon. However, it is a valuable option to have if the value of the project drops after construction due to an economic downturn or due to other factors.

2. MaxCon can abandon the project any time during construction and incur no liability. Although no one is hoping for this outcome (as MaxCon would have lost all investment until abandonment), these are also valuable options to have. The project leader knew that these options cannot really be exercised any time as the company must commit to suppliers and building contractors for a period of time. Yearly contracts are the norm, so these options are potentially exercisable at the beginning of each year.

3. The penalty set forth for any delay in the project is not an option but something MaxCon must do if the project is not completed within the project timeline (4 years).

4. Every year during the operating phase, MaxCon has downside support at $7 million. If its debt service is higher (selecting favorable currency), then the support increases to the debt service payment. In effect, it has the option to elect to receive the maximum of the tolls collected, $7 million, and debt service payment in favorable currency.

5. Finally, the project has a terminal value as a function of tolls collected in the last year of the operating phase. The higher the last year's tolls are, the higher the terminal value will be.

Figure 12.12 shows the model. The value of the project is $17 million. If MaxCon bids the project at $17 million, it will not make any money, and the expected return is zero. If it bids more than that, shareholder value is destroyed. A bid less than $17 million will add to shareholder value. After much discussion, MaxCon settled on an initial bid of $7 million. Success will increase shareholder value of the company by $10 million.

The risk-neutral payoff distribution is shown in Figure 12.13. Provided in Figure 12.14 is the cumulative probability distribution of risk-neutral payoff.

The following are worth noting:

1. The project is risky. The downside is limited to a loss of $40 million, and the upside is as high as $100 million. Without government guarantees, it is unlikely that any commercial enterprise will be interested in a project like this one.

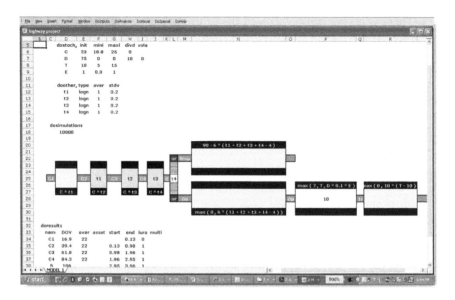

FIGURE 12.12
Build, operate, and transfer project model.

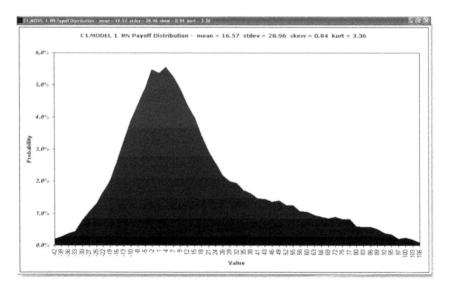

FIGURE 12.13
Risk-neutral payoff from BOT project.

2. There is a 33% chance that the company will lose money in the project in spite of all the features, such as the ability to sell back, as well as all the various guarantees by the government.

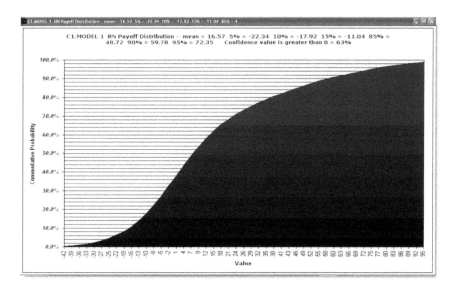

FIGURE 12.14
Cumulative probability of risk-neutral payoff from BOT project.

3. This project is a good candidate for winner's regret as a slight increase in the winning bid can result in a loss for the winner. In a bidding situation, it is more likely that companies overbid. Understanding the economic value of the entire project, taking into account all associated input uncertainties as well as decision flexibility embedded in contract features, is essential for companies both to bid successfully and to win projects that add value.

13

Misperceptions Regarding Real Options

There are debates among academics as well as practitioners about the usefulness of real options techniques in investment analysis. In the past couple of years, we have seen no fewer than 20 major conferences focusing on real options. And, while it appears that the interest in real options techniques is increasing, the debate goes on, and there continues to be a great deal of skepticism.

Options thinking is important when one deals with uncertainty and decision flexibility. In situations with little variability in expected outcomes and no flexibility in future decision choices, conventional techniques like discounted cash flow (DCF) are adequate. But, once some of the assumptions that underlie the traditional techniques are no longer applicable, it makes sense to move toward a broader options framework. There are classes of decisions in every industry that require a method like decision options analysis, but it would be a mistake to move into a broader framework when the assumptions underlying traditional techniques are valid. In other words, complicating the analysis unnecessarily does not add value. There also appears to be a lot of confusion about how to represent the various types of risks in the real options methodology. There are two major types of risks we consider in all financial analysis: private risks and market risks. An example of a private risk is the possibility that toxic effects are revealed in a clinical experiment on a pharmaceutical drug that forces the abandonment of the program. The likelihood of a decline in sales for lifestyle drugs if a recession hits the economy can be considered market risk.

One common argument against the use of real options techniques is that they do not work when private risks dominate market risks. It is true that when the value of an investment is heavily dominated by private risks, the managers of the investing company have much less decision flexibility. In some sense, when the investment opportunity is driven by private risks (sometimes called unique risks or technical risks), the decisions are made by acts of God or other such phenomena over which the decision makers have no control. And, it is also true that when there is little flexibility, traditional techniques will provide an adequate way of estimating value. But, the point to keep in mind here is that private risks are treated the same whether you use real options or the old DCF approach. Only the market risks that are treated differently.

Some question the empirical validity of the real options approach. The problem is that it is difficult to empirically validate a decision methodology. In private markets, the real options approach helps improve decision quality. Decision quality means only that all information is used systematically in making a decision. Because there are many private risks in a decision (such as technical failures in a research and development [R&D] program in the pharmaceutical industry), we cannot validate the decision after the fact. That is, the success or failure of the R&D program, on which the decision is made, does not tell us if the decision was a good one. The decision maker was already aware of the probability of a technical failure. The fact that the technical failure happened does not mean the decision maker was at fault. Similarly, one can also question whether we have proof that markets are pricing assets by simply discounting cash flows according to the capital asset pricing model (CAPM). Many studies showed that only a small percentage of the total value of certain types of companies (such as so-called information and knowledge firms) can be explained by DCF values. Much of the rest of the value comes from the so-called growth options or investment opportunities. For pharmaceutical companies, this growth option value represents as much as 70% of market capitalization. There is no doubt that the market is capable of appreciating value beyond what can be assessed using traditional DCF techniques. For example, many of the early users of the Black-Scholes formula for stock options, eager to make money with the new technique, came to the painful conclusion that markets were already pricing options in essentially the same way as proposed by the mathematical equation (prior to the existence of the equation). What the Black-Scholes formula did was to model an existing market pricing process and thereby institutionalize it. In similar fashion, real options analysis should be viewed as providing a framework—one that is lacking in traditional corporate finance—that replicates and quantifies sound management intuition about the kinds of investments that are most likely to create value over the longer term.

Some companies fear that if they adopt real options in decision making, their stock prices will suffer. Managers of these companies tend to believe that markets are pricing assets using a highly simplistic DCF model—one that effectively capitalizes only next quarter's earnings (or at the most, next few quarters' projected earnings). But, to the extent we can judge from the high P/E ratios of many growth firms and from the willingness of investors to assign large values even to companies without earnings, this view clearly seems mistaken. Based on the current concerns about corporate governance, those managers who take shortsighted actions to boost quarterly earnings per share (EPS; in an attempt apparently to boost stock price) are most likely to find themselves punished by the market. Next quarter's earnings are important, to be sure, but not as important as the company's ability to demonstrate that it has a coherent strategy based on building and maintaining its long-running competitive advantages. To the extent conventional capital budgeting techniques are preventing managers from thinking strategically and encouraging them to use tactics for meeting near-term earnings targets (originating from a simplistic net present value [NPV] model), the long-term impact on the company will certainly be negative. Managers need to adopt decision options thinking and, equally important, find effective ways to communicate their thinking to help investors understand the value of the company's portfolio of growth options and opportunities. Some point to the negative reaction in stock prices to an earnings miss by a company as evidence that market uses discounting of quarterly earnings to estimate the value of equity. It is important to note that the negative reaction is likely far in excess of what can be predicted by the difference in actual and predicted earnings. If the company has conditioned the market to expect a precise EPS, through estimates and projections, the market may take the management on face value. If then the company misses the highly publicized quarterly EPS, the market will likely punish the company for two important reasons. The market may put an "incompetence discount" on the company's management as unable to even perform proper accounting and forecasting. More important, the market may also see a possible signal in the company's inability to meet targets, indicating that the future growth potential of the company is declining. In either case, the decline in stock price will be far in excess of what can be calculated by simple discounting.

There is also a perception that real options are too complicated to be practical. We can blame some of the academic research in this area for creating such a perception in the minds of decision makers. It is certainly true that Texas Instruments does not market a real options calculator (yet). In this book, we solved complex problems using the Decision Options software. There are also complicated calculations in engineering, with detailed modeling of aircraft parts relying on techniques such as finite element analysis and computational fluid dynamics that are certainly not available on a calculator or in Excel. Yet, large aircraft manufacturers do not shy away from these techniques. Corporate investment decisions should be evaluated using the best available techniques. An accurate assessment of value can make a difference, and a technique that provides a more accurate quantification using all available information should not be ignored. Of course, an analysis does not have to be complex to be correct, but perceived complexity in a technique is not sufficient grounds for dismissing it. There are many ways to make real options models more user friendly, but such models will never become as generic as an Excel formula because decision options techniques require careful framing of the problem, beyond a forecast of expected cash flows followed by rote discounting.

I do not know of any good senior executive who would approach an investment problem as a now-or-never proposition with no future flexibility or variability (the assumptions one is implicitly making in a DCF analysis). It would be hard to find a senior executive who, faced with a 5-year investment program, does not think about delaying, abandoning, or expanding aspects of the program over time, investing on a small scale initially as a means of learning about cash flow potential, or forming contingency plans given a very uncertain future. This is, and has long been, a natural way of thinking for managers. Until real options were developed, however, corporate finance had not provided a way to structure such thinking and quantify the value of strategic alternatives. Try explaining DCF to a good senior executive. It seems unnatural to forecast single-point estimates of all cash flows and then discount those back at one rate (typically the weighted average cost of capital) to get an NPV. More sophisticated finance departments run scenario analyses to create optimistic, normal, and pessimistic projections. But, it is also important to keep in mind that neither scenario analyses nor Monte Carlo simulations of NPV are good substitutes for real options analysis since neither incorporates management flexibility—and there still is the problem of how to determine an appropriate discount rate. To assume that there is no flexibility in the future and that one needs to make the investment now or never is simply unrealistic. In more ways than one, good senior executives have internalized real options thinking, which makes it more difficult for them to be satisfied with standard DCF analysis. And, in many cases, they make decisions in spite of the financial analysis, which can be a frustrating experience for financial professionals starting out in the corporate world, newly armed with DCF and NPV.

There is also a perception that the data needs for conducting real options analysis are very high compared to traditional techniques. The amount of data used in an analysis is more a function of the analyst than the method employed. A chief executive officer of a Fortune 100 company once said that he was more comfortable with one data point than with two—because he could draw an infinite number of lines through one, but with two data points he felt constrained. In general, real options analyses actually tend to be less data intensive. Since the average underlying value is used together with an estimate of volatility (variability), real options users tend to be comfortable with less precision in that average value. But in a traditional DCF analysis, all the information is wrapped into the average cash flow, so there is a stronger focus on getting the average right. As a result, extensive market research is conducted and scenario analyses are performed to fully define the average. Curiously, after exerting significant effort in collecting the data and performing these analyses, most of the information is thrown away, and the expected (or average) cash flows are then discounted at the weighted average cost of capital to create a generally meaningless NPV.

It is true that the real options analysis takes a bit more effort to fully appreciate. On the other hand, how many decision makers really understand the underlying assumptions of the CAPM when they use DCF? It is not necessary that all users of a technique understand all the underlying assumptions and mathematical derivations. In fact, decision makers routinely take NPV at its face value without fully appreciating the math underlying it. Moreover, the prescribed mechanics and assumptions of the CAPM are often violated in practice, but nobody seems to care much.

Consulting firms that promote decision tree approaches have long maintained that they are performing real options analysis. The problem here is obvious. Decision trees are merely pictorial representations of DCF calculations with private risks, represented in the branches. The appropriate probabilities are applied to the cash flows past that decision point, just as we adjust the cash flows for risk before discounting them. In performing DCF, we adjust (or should be adjusting) cash flows in exactly the same way. Decision trees are a good way to frame the decision problem and are good communication vehicles, but they do not provide the benefits of a real options analysis.

Some argue that real options techniques simply increase value to make a project more attractive. It is true that a correctly formulated DCF analysis will typically give a lower value than its real options cousin. However, it is important to keep in mind the role of the discount rate as a key driver of value in DCF analysis. As mentioned, discount rates are not generally observable in the market for a specific project or decision. For this reason, managers who are predisposed to taking a certain project (regardless of its value) are able to start with discounting at weighted average cost of capital (WACC) and then adjusting it downward until they get a believable number for the value of the project. Although people like to say that valuation is more an art than a science, I contend that there is (or should be) a lot more science in valuation, and that assumptions that are not market based make the whole process useless.

I once had an experience with an investment bank that valued a biotechnology company using traditional techniques. The bank performed a bottom-up valuation of the company's products in the R&D phase, but failed to adjust the cash flows for private risk applicable to R&D. It then calculated a figure that was triple the current stock price of the biotechnology company. A real options analysis resulted in a price much closer to the market price as framing the problem with explicit separation of private and market risks make it less likely that some risks will be "forgotten." I suspect that the ease of use of the DCF technique in Excel and other such analytical tools may have let these types of mistakes creep in.

Real options got some of the blame for the technology bubble (artificially inflating prices) and the failure of Enron. The bubble arose from a technology discontinuity. Bubbles form with or without real options (as we have proved in the U.S. real estate market recently). The Enron failure resulted from fraud. In fact, if real options analysis had been more mainstream in late 1990s, it might have prevented the technology bubble that we experienced. Market participants may have had difficulty assessing value in times of high uncertainty without sophisticated tools such as real options. A reliance on rules of thumb and precedent may have actually contributed to the overvaluation of the technology stocks, which tend to show extremely high volatility. The technology bubble is a strong argument for, rather than against, the institutionalization of real options analysis in valuation. Regarding Enron, I think the perpetuators are enjoying accommodations in a federal penitentiary, unfortunately without any options.

The bottom line is that the real options framework is only a generalized asset-pricing model. Every industry encounters situations that could benefit from such a broad methodology. DCF is nothing more than a special case that is applicable when there is little variability in outcomes and no management flexibility in decisions. But, in the absence of these simplifying assumptions, real options techniques are likely to provide a more complete financial analysis for the decision maker. In fact, good managers have always thought and acted in a real options way—in large part because markets appear to value companies and products in the same way. It is important for companies, particularly those for which "growth options" constitute a significant portion of their value, to institutionalize analyses based on decision options in their decision processes. Those companies that use traditional techniques and focus excessive attention on near-term earnings in their investment decisions are likely to be punished by the market in the long run.

14

Challenges to Decision Options Implementation and Possible Remedies

As you set out to solve decision problems along the lines discussed in this book, you are likely to encounter several challenges from both academic quarters and institutional resistance to change. Let me list the most likely arguments that one is likely to encounter against the use of options thinking in decision making and possible ways to overcome them.

First, it is my belief that academic research in the area of real options since 1990 is primarily responsible for the perception that exists in the industry regarding the complexity and applicability of options thinking in decisions. Academic research has centered on highly simplified and stylized examples of real options, many of which have no relevance for practical decision making. However, the solutions of these "simple problems" tend to use a lot of high level mathematics and intimidating long equations in an attempt to find closed-form solutions. It is also the case that recent academic publications focused on advanced topics such as "game theoretic applications of real options." This has further reinforced the idea that real options are something academics love and have no relevance for practical decision making. Moving into highly advanced topics such as these before simpler decision approaches are implemented is dangerous, especially if publications attempt to push these ideas into practice. There are many academic challenges to the application of real options, and much still has to be done to create a more robust framework for analysis. However, there is a distinct risk that academic research in this area will become disconnected from practice, and as progress is made in the abstract analysis of the concept, the practice of real options will stagnate. I hope the ideas discussed in this book will initiate a more constructive dialogue between theoretical research and tangible applications.

The first academic objection to options-based valuation of real assets is that the underlying asset is not "replicatable," and hence the "no arbitrage argument" that we invoke to value the derivative in a risk-neutral framework is not valid. This, indeed, is true in many of the cases discussed. Every time I have presented these ideas in conferences, someone in the audience points this out. He or she will argue that the asset (such as a drug candidate) on which I valued the option (such as the option to conduct a clinical study) is a private asset, and there is no proxy for it in the marketplace. Thus, the risk-neutral valuation I employed is simply not valid. When asked what alternative methodologies they may use, most say, "I don't really care; all I can say is that you cannot do what you describe here because the asset is not replicatable."

In essence, what the person is saying is that markets are not complete. In a world where technology and location flexibility are increasing, companies with larger portfolios (built by legacy scale and scope advantages) will continue to disintegrate into smaller units, providing a richer array of pure companies and entities, warming the hearts of those yearning for market completeness. Meanwhile, financial innovation will continue to create instruments with increasing focus, such as the exchange traded fund for the disease of depression (using the stock of companies that focus in this area), moving us closer to "perfect replication" of the underlying. We can look forward to technologies, organizational structures, business processes, and financial instruments in the future that will help us sleep better knowing that the world is moving toward market completeness. However, markets are never complete (and it is the case for traditional finance also), so we will have "return shortfall" in practice from the pure case. Meanwhile, we know that uncertainty and flexibility are important considerations in decisions, and it may be better to include them than pretend that they do not exist.

However, if the person is objecting to the use of options pricing as a method, arguing that traditional finance (discounted cash flow [DCF] based on the capital asset pricing model [CAPM]) is better, then we may have stronger arguments. Application of CAPM DCF is also not perfect as it will require the user to determine a discount rate based on the correlation of the asset's return against the return of a market portfolio return or to find a close enough proxy. Since proxies for the asset under consideration and the market are hard to find, often analysts fall back on "easier ways." These may include the use of discount rates based on WACC (weighted average cost of capital) or corporate prescriptions of a constant discount rate. In some cases, round numbers such as 10% are used to discount cash flows. Excel has made cash flow discounting easy, and this has provided confidence to the participants. In a high percentage of the cases, the net present value (NPV) determined from discounting deterministic future cash flow estimates with an arbitrary discount rate is not useful for decisions, and this practice is not valid according to theory.

In decision options pricing, however, we have the following advantages:

1. Uncertainty and decision flexibility are considered systematically. In most decision situations, the factors that drive value such as revenues, costs, timelines, and the like are highly uncertain. It is also the case that decisions are complex, and typically future decisions, which are contingent on the present one, exist. Decisions may also provide alternatives (how), timing flexibility (when), show options characteristics (right to do something but not an obligation), interactions,

and path dependencies. These are rich attributes of the decision and cannot be ignored. One of the reasons good decision makers are compensated well is that, in the absence of tools, shareholders do not have any alternatives but to hire the best "gut-based" decision maker, who in the past has shown "good judgment." Note that part of this good judgment is the ability to envision and consider flexibility, and the other is the ability to estimate uncertainty—both private and market—so that one can make the best decision considering all risks. Some reach here by trial and error, but errors in business can be quite expensive. Some are simply very good at making decisions. However, if companies have a method and tool that allow a systematic replication of this gut feel and good judgment, they are better off. They can use the method in all decisions, regardless of personalities, and can train upcoming managers in decision making without the more expensive trial-and-error. Although the decision options method is not perfect, it is far superior to other alternatives.

2. The use of a risk-free rate (in risk-neutral valuation) removes "gaming" of analytics. This is a major issue for the practice of traditional NPV-based decisions. Since there is no systematic way to determine an appropriate discount rate, the rate can always be changed if the answer is not what one expected. This happens in a large number of decision situations, which makes one wonder what such a valuation exercise is really accomplishing. The risk-free rate is observable in the marketplace, and as you have seen in many of the cases described in this book, a real risk-free rate is generally close to zero. Use of a zero discount rate and real cash flows allow us to be consistent in our assumptions and methodology. Since private risks are captured explicitly (and not lumped in with the discount rate), this practice gets us as close as we can to the current theoretical understanding in economics (it is not perfect, but neither are other alternatives).

3. Just as in the correct application of CAPM, in which private risks are used to adjust cash flows and market risks are used to determine a discount rate, decision options also allow an explicit split between private and market risks. In decision options, as seen in the cases discussed, the market risks can be disaggregated and used with varying characteristics (such as mean reversion and jumps), and private risks can be represented in various ways (such as binary and probability distributions), providing a much more complete representation of the decision problem. It is not perfect but much more complete than the current alternatives.

Applications of stylized theory and framework will always be imperfect, so perhaps it is acceptable to use the options methodology even with its apparent academic deficiencies.

The second academic argument is that markets are not efficient. For tradable assets, one cannot really assume that the price process follows random walk. Remember that in many of the cases discussed, we assumed that the underlying asset's price followed random walk or geometric Brownian motion. If markets are not (at least) weak-form efficient, it would mean that past prices may have an effect on future prices, and the price process cannot be assumed to follow random walk. It is clear that billions of dollars are spent on Wall Street to prove or disprove tomorrow's forecasts based on yesterday's information. Fortunately, there are still some academics who believe in the abstract concept of efficient markets. From a more practical perspective, it is easier to deflect this by pointing out that technical analysts on Wall Street are still going to their jobs every Monday morning. If the "head-and-shoulders" chart patterns do work, all technicians would have climbed to the list of the wealthiest in the world. Fortunately for the proponents of efficient markets, this has not (yet) happened. It should be noted that the idea of efficient markets is a theoretical construct. There are many markets and situations for which the degree of efficiency can differ. The primary takeaways from this elegant concept are two-fold:

1. In a world of fast-moving information and technology, prices of assets adjust to available information fairly quickly. The price of a marketed asset is what is observed in the marketplace, even if one has doubts about it. Note that we are excluding fraud and assume that the information available to the market is true. It is hoped that the group of white-collar crooks, currently enjoying accommodations in jails for financial fraud, will help us assume that the available information is generally good.

2. If prices adjust to new information, past prices are not going to be useful in predicting future prices. Arrival of information in the past and subsequent adjustment of prices to that information do not tell us anything about when and what new information may arrive in the future. If this is not the case, wealth will accumulate for some at a really fast rate as they successfully predict stock market moves.

Efficient markets mean that prices adjust to new information (reasonably quickly), and past information is already reflected in the current price. It is difficult to object to this in practice. If we can convince the disbelievers of this somewhat simpler articulation of efficient markets, we are in the business of assuming prices of traded assets follow random walk.

The third argument, although less academic than the previous two, is that real options are not optimally exercised by managers. In the valuation process, we assume that the decision makers are rational, and they optimally exercise, sell, or abandon their options. Some practical-minded academics have pointed out that managers never do this. Misaligned incentives, a localized need to save jobs, building bigger departments, and the ego may prevent managers from making optimal exercise decisions. Those who invoke this argument ask, if managers do not exercise options optimally, how can one determine economic value by assuming that they do? They argue that such valuations and attempts at making decisions more optimal constitute a waste of time as long as managers behave as human beings, always interested in maximizing personal wealth rather than shareholder value. This is indeed true. If companies are mired in agency problems and managers exist to destroy value (by suboptimally exercising options), it is not really useful to conduct any analysis, let alone options analysis. It is hoped that the owners eventually catch up with managers with a chronic habit of suboptimal options exercising and remove them from their jobs. It is more of an agency problem in the enterprise, and it will be difficult to wait to think about uncertainty and flexibility until all incompetent managers are removed from their jobs.

Even if one is able to win over the academics to the idea of options thinking, despite its academic impurity, it is in fact a better way to make decisions in companies even if further implementation difficulties arise.

DCF-based NPV (and internal rate of return [IRR]) and decision trees have been ruling the financial decision-making process in companies for over 30 years. Many companies have invested heavily in tools, accounting, and business processes that utilize traditional finance. For example, decision tree-based analysis is prevalent in pharmaceutical companies. Consulting firms have been helping these companies make project selection, design, and portfolio management decisions for over two decades based on traditional decision trees. Because of the long research and development (R&D) cycles, there are no feedbacks to these decisions, and no one has asked if such methodologies have actually helped these companies. This is a slippery slope as traditionalists may ask options advocates what evidence they have to argue that options analysis is any better. The best way to win this argument is by suggesting that we know uncertainty and flexibility exist in R&D, and it may be better to take account of it rather than ignore it.

A few consulting companies, while making "strategic decisions" for their clients, have argued that decision trees indeed represent options analysis. They argue that options are considered at all decision nodes of a decision tree, and by articulating the various choices that exist in the future, decision trees capture the essence of options. They also say that the elaborate decision trees that exist for R&D programs alone demonstrate the systematic application of options analysis. What is unclear is whether such decision making has enhanced shareholder value. This has bred a bit of confusion in the enterprise. The two important considerations of options, uncertainty and flexibility, are not present in decision tree analysis (DTA). Decision trees are pictorial representations of the DCF mechanics. The branches in a DTA represent private risk (similar to how cash flows are probability adjusted in the numerator in a DCF equation), and the NPV that is produced is a blind discounting of the entire tree (using a generally arbitrary discount rate) without consideration for decision flexibility.

The DTA advocates may argue that to capture uncertainty, they do scenario analysis; so instead of one NPV, they create three or even more NPVs (such as average, optimistic, pessimistic, and really pessimistic scenarios). In this case, DTA analysis takes in apparently precise inputs (such as cost, timeline, probability of success, and peak sales) and creates uncertain outputs (such as average NPV, optimistic NPV, and pessimistic NPV). One may want to suggest that it may be better to consider uncertainty in inputs (which we know exist) and create a single output (rather than many possible outputs) so that one can make a decision and act on it. For example, if one is trying to buy or sell an intellectual property (IP) position, having three or more different NPVs does not help as the checkbook can take only one number.

Regarding the apparent capture of options by the branches of a decision tree, one may want to suggest that if NPV is calculated (in the DTA) by assuming all future decisions are made now (and they are not contingent), that does not capture flexibility. If future decisions are choices or options, such decisions will be made in the future and not now. The branches of a decision tree are not options but a representation of private risks. Also, use of a constant discount rate across investment opportunities is fundamentally at odds with the application of the CAPM itself. So, DTA has no market uncertainty and no decision flexibility, and the discount rates used typically are WACC or 10% (for ease of use), firmly removing any semblance to traditional finance theory (including the CAPM).

Even if one survives the wrath of decision tree consultants in attempting to impart options thinking in their companies, there may be further obstacles. One issue is that (at least in the United States) decision makers have gotten too used to IRR and DCF analysis. The calculations are very simple, and everybody understands them. Options analysis is not that simple, and this creates anxiety for decision makers. Managers are not comfortable making decisions on attributes or calculations they do not understand. If better tools exist, this may not be an issue, and we are making progress in this dimension. It will be helpful to get the decision maker focused on the assumptions (and the uncertainty around the inputs) as well as the decision framework (and associated decision flexibility) rather than the output. If confidence can be built that newer tools have made the calculation mechanic facile and fast, it will allow the decision makers to engage in richer discussions around the factors that drive the decision. Such a process in which analysts and decision makers engage in describing all uncertainty in assumptions as well as choices they can make in the future will help to increase both confidence and decision quality. The lack of confidence and the lack of comfort in results from traditional analyses force them to commission more analyses, leading to many results—average, pessimistic, optimistic, and so on—making clean decision making difficult, if not impossible.

Other issues are executive compensation and managerial incentives. In the past two decades, companies have moved into using employee stock options as a major incentive mechanism for key employees. However, managerial decisions have asymmetric payoffs for the manager who is making the specific decision because decisions have differing impacts on the individual and the entire company. Even if the manager believes a better method exists to reach certain decisions, the manager may not commission it. This points to a larger agency problem in the enterprise, and it requires more attention from the shareholders and organizational theorists. As tools improve in analyzing information systematically and creating more optimal decisions, it is imperative that companies design appropriate organizational structures and incentive systems to take advantage of them.

One other form of institutional resistance comes from the decision makers themselves. Some are not comfortable with the "systematization" of the decision process they have been conducting qualitatively for decades. If good intuition can be replicated, some may think that it will undervalue what they currently do. After all, those who make good decisions on a gut feel are sought by shareholders, so those who are good (or think they are good) have no reason to implement any technology in the decision process. Even if one is good in selecting and designing projects and managing a portfolio of complex projects by gut feel, he or she has to consider obligations to the shareholders of the company. What if this person can improve his or her intuition or, even if he or she is perfect, still has to develop future decision makers for the company? Can this person really hope that the next chief executive officer will be as good, or will the new person require a series of trials and errors to improve? In the latter case, the person should consider more systematic approaches to decision making as the future of the company may depend on it.

Academic and institutional challenges exist in the implementation of options thinking. Academic challenges to options analysis can be overcome by pointing out that markets are reasonably (not perfectly) efficient, markets are increasingly (not fully) complete, and managers are reasonable (not fully rational) human beings; all theories provide a framework for thinking, and practical applications will always be imperfect. Decision options is a generalized method that takes into account technical and market uncertainty as well as decision flexibility. It is better to consider uncertainty and flexibility that we know exist rather than ignoring them as is the case in traditional finance.

The institutional resistance to change is sometimes more difficult to overcome because certain participants have a vested interest in perpetuating status quo or decision makers are uncomfortable or unwilling to consider new methodologies. Here, we need new and better tools that will make the analysis more facile and faster, allowing decision makers to focus on what is important and avoiding the somewhat meaningless debates around uncertain results. If organizations are designed better with more aligned incentives for managers, it will become easier to more broadly apply options thinking in decisions.

In summary, decision makers should not wait until markets are complete and fully efficient to start incorporating options-based analysis. They also may not have to wait to think about these ideas until all incompetent managers (who do not exercise options optimally) are removed from companies by frustrated owners. Also, uncertainty (private and public) and flexibility are integral to options, and any theory, methodology, tool, or scheme that does not take these into account systematically is not options analysis or applicable across investment choices that exhibit these characteristics. These ideas have been with us for over two decades—the academic debates and decision tree consultants have prevented companies from taking advantage of them.

15

Past, Present, and Future

The Industrial Revolution ushered in an era of a production economy in which a single powerful idea resulted in gigantic industries and huge enterprises that thrived on scale. In the scale economy, efficiency was very important, and managers had to focus on getting the conveyor belt moving in the most optimal fashion. Hand-and-motion studies were conducted to assess a better way of producing goods from a generally unchanging specification. Strict time records were kept, and wages were only functions of duration of work or production. The contract between the employee and owner was simple: wages for time worked or goods produced. Measurement was reasonably straightforward: recorded time on punched cards or paper or a count of produced units of goods. This was sufficient to execute the wages part of the contract. Work was characterized as a commodity product—a substitute could be found with relative ease. This was an era of determinism.

The introduction of a manager in the scale economy meant a slightly higher specialization in work. Some of the skills of the scale economy manager—a strict adherence to rules, error-free record keeping, and the handling of money—were different from the skills of a production worker. These skills made the manager less of a commodity, although substitutes can be found even for that skill set. This is the beginning of the skills-based class structure in large enterprises. When the owner noticed such a difference (as did the manager), the contract between them became a bit more complicated than the worker–owner contract. Ease of measurement of work done and the linear relationship between compensation and work existed in contracts. One of the differences in the manager's job, however, was the criticality. The downside risk of underperformance of the manager was higher (from the owner's standpoint). The owner thus was willing to take insurance against such downside risk through a positive reinforcement of an attendance bonus or a negative reinforcement, such as a threat of job loss for absenteeism, the former representing a call option for the manager and the latter a put option for the owner. A manager's base compensation reflected the existence of these options over and above the base. As complexity of production (such as skills specialization in workers) increased, the manager's role began to take a higher dominance in the overall system. The manager (because of intimate knowledge of workers) held information such as individual worker productivity and the proclivity toward absenteeism that allowed the owner to optimize production. This upside potential further drove the owner to enter into complex contracts with the managers that included a production-based bonus and profit sharing. Such contracts began to show complex embedded options for both parties to manage uncertainty. All these trends increased the managers' power in the enterprise. Quality of management was then measured by the manager's ability to produce more work (according to relatively stable specifications and simple instructions).

As economic wealth increased, industries matured, and consumers became more knowledgeable, a new competitive arena, differentiation, became feasible. In this environment, value became a function of features and utility in addition to weight and numbers. In this environment, the power of a single idea diminished, and integration of a sequence of ideas began to command higher competitive leverage. These ideas were generally extensions of an existing dominant idea or the bundling of one or more existing ideas. Since the owner's value was now a function of production scale and scope of products, the owner–manager contract needed to be constructed to consider both aspects. Overall profitability of the owner was dependent not only on aggregate production and efficiency but also on the rate of improvements made to the original idea. The value of the improvement was a function of both timing (how quick) and relevance (how profitable).

The initial reaction of the owner to the changing environment was the introduction of higher specialization of workers. Production and design departments were created and kept apart to ensure clean lines of accountability. Manager specialization was increasing as well. Production and design managers with different skills were sought even though the fundamental role of either manager was about the same: higher production. The production worker and manager contracts were relatively the same in the new environment. For the design worker and manager, the measurement complexity was higher. Since both the timing and the relevance of the specification change were important for owner profitability, incentives had to be designed to optimize both. These changes in production specifications, however, led to higher inefficiencies in production, and the scale advantages (on which the whole enterprise was based) were slowly disappearing for the owner (and to the owner's dismay for the production manager as well). If the worker had a strict quantity contract, his or her compensation declined (as a changed specification took more time to produce and the brain needed to be retrained for a different set of routine activities).

This incremental disutility was compensated by the appearance of variety for the production worker for the first time. The possibility of replacing a set of routine activities by another similar set of routine activities at regular intervals added utility to at least a cohort of production workers. For the production manager, this posed new challenges. Measurement of worker productivity became more complex, and calculation of compensation from quantity and time worked were not enough. Worker productivity did not follow a linear relationship with experience (some were more adept at changing quickly than producing more of the same specification). With these challenges came new information to the production manager that could be used to leverage profitably for the owner. Both production worker and manager contracts changed to take into account, for the first time, the ability to change quickly into producing an incrementally different specification drawn by the design department. Technological change during this time also lent a major blow to the scale-based enterprise. The idea of higher efficiency through increased scale was challenged by the ability to achieve rapid customization through technology.

There was more tension looming in the horizon for enterprises in the scope economy: a natural tension between the design and production departments. Complexity in contracts led to misaligned incentives between silos. This led to the design department changing specification with no regard to production difficulties and the related erosion of profits for the owner. Forward-thinking owners attempted to combine the departments or align incentives more closely to overall profits. Such owners quickly realized that design and production were not independent activities, and that production was likely to be the best place to extract design ideas. Leveraging the worker knowledge (now fashionably called worker empowerment) helped many companies increase productivity. In all this chaos, managers gained more power as the owners generally lacked information to reach conclusions on optimal structures. Unfortunately, the managers' utility curve had competing priorities. One of these is the desire to have a large group of people working for the manager. This was partly the result of incentives in the contracts that the owner imposed on the manager. The power base was a function of the number of workers as the owners' expectation of the managers' knowledge was highly correlated with the number of workers reporting to the manager. This scale mentality led the owner to make incorrect decisions, and the owner was generally unable to maximize value.

The spread of ownership through mature equity markets and the efficiency of information flow finally provided a new way out of this manager–owner stalemate. One outcome was the removal of managers whose utility curves were in conflict with overall firm value. To the surprise of owners, the scale and scope economy had created a huge growth in the number of managers, and the removal of a substantial part of this management layer actually increased firm profitability. This manager downsizing was also accompanied by the breaking up of large enterprises into smaller entrepreneurial units with higher alignment of incentives among workers, managers, and owners. This increased flexibility and the ability to manage the enterprise under uncertainty. This brought huge incremental value to the overall economy.

During this time, the business environment was experiencing the third major change, one that is driven by innovation. Industry finally arrived at a point in time that recognized that the brain is in fact more powerful than muscle. Scale from the production of powerful single ideas and the incremental extensions of such ideas through scope gave way to the fundamental need for innovation. This was the first real break from the ideas that have been with us for so long from the Industrial Revolution. The innovation economy is fundamentally different from the previous situations. In this environment, there was little role for the production worker or manager who performed essentially routine activities. Technology was available for automation of the routine, and leveraging the skills of the workers into innovation was the most important competitive lever. The transition into the innovation economy provided opportunities for companies to configure their value chains, taking each other's strengths and weaknesses into account.

In the innovation economy, shortening product life cycles, diminishing the economics of product extensions, and accelerating the wealth of demanding and discriminating consumers all point to a very different way of doing business. The competitive advantage is solely dependent on how fast and how profitably new ideas are generated in the new environment. It is not the quantity of routine production but the quantity of essentially nonroutine innovation that is important. In this new era that is driven primarily by the rate of innovation, the traditional management processes and techniques need to be changed. The contract between owners and managers needs to be redesigned, and the incentive system needs to be rethought.

One important idea in this transformation process is flexibility. In the scale-and-scope economy driven by efficiency, we strive to reduce flexibility as flexibility generally hinders orderly production. However, if innovation is the driving factor, flexibility has to take the front seat. Flexibility has to be used as a fundamental guiding principle—in organizational design, incentive and contract designs, human resources, infrastructure, planning and scheduling, and capital structure—in every aspect of business. What does flexibility mean in practice? Business thinkers have been using the term *organizational agility* to describe the ability of the organization to quickly adapt to changing environments, competition, and technology convergence. Flexibility is an organizing principle to enhance agility. Flexibility allows an organization to both identify trends early and adapt successfully. In general, as the size of an organization increases, flexibility decreases. One way to stay flexible is to design the organization in smaller entrepreneurial units that are tied together not by rules but by culture and competence.

Human resources is treated as an afterthought in many companies, and the term generally implies people management and not value enhancement. In the innovation economy, human resources assumes critical importance, and its treatment as a commodity in the scale-and-scope economy needs to be abandoned. Innovation is a result of both flashes of brilliance of individual creativity and systematic problem solving in groups of varied competencies, skills, and expertise. Hiring based on set specifications and offices that resemble jail cells no longer work. Introducing higher levels of flexibility in human resources with a variety of skills and expertise and designing a work style that draws in more unconventional participants (who may not adhere to the standard 9–5 rules) are important for the development of a workforce capable of innovating at an accelerating rate. Companies also have to realize that diversity is an important ingredient for flexibility. Human resource policies that advertently or inadvertently encourage similarity in thinking, education, location, and physical characteristics will reduce flexibility and make the company vulnerable to failure in uncertain times.

Incentive and contract designs are also increasingly important in the innovation economy. For example, configuration of the value chain of the company, including its suppliers and buyers, needs to internalize a systemwide flexibility: redundancy in critical suppliers; ability to increase and decrease purchases as new information is revealed; ability to reroute and reconfigure in the event of major disruptions; ability to switch products, locations, and processes; and so on. Such systemwide flexibility needs to be reflected in better contract designs between partners that incorporate flexibility systematically and price them efficiently. Design of incentives has shown little innovation in recent times as most companies follow set principles such as stock options and top-down pyramidal compensation schemes inside the company and rigid contracts with suppliers and buyers outside. None of these have shown to be particularly effective in increasing productivity and shareholder value. One of the primary reasons is that these instruments are designed more for the scale-and-scope economy, and they generally fail in the innovation era. New instruments and methods need to be designed that will enhance innovation by the creation, nourishment, and optimal exercise of real options. Incentives should encourage better management of uncertainty rather than attempts at eliminating uncertainty. Decision makers have to understand that uncertainty is a source of value if managed with flexibility, and conventional notions of minimizing risk at any cost will lead them down the path to mediocrity.

Corporate finance also has been stagnant in recent years in many aspects of capital structure, project financing, and portfolio management. The general tendency has been to apply traditional theory and assume that the capital providers also follow such ideas. For example, a significant amount of time and effort of senior decision makers is wasted in the "management" and "communication" of quarterly earnings. Although empirical evidence has been strong that enterprise value is a function of the future prospects of the company and the ability of its management to create, nourish, and optimally exercise options, many still act as if value is driven by a blind discounting of earnings of this quarter and projected earnings of next quarters.

Companies have to depart from traditional notions that real asset decisions can be fully disconnected from the financing decisions and create a holistic portfolio management process that includes both. Management of the company in silos—where research and development (R&D) decisions are made disconnected from financing decisions and capital structure decisions are made in the absence of a good understanding of the risk of the real asset portfolio—is fraught with danger. It is clear that the "local optimization" attempted by large companies—in providing incentives for R&D to increase throughput in product extensions, marketing to increase volume, and finance to meet the numbers—is by definition suboptimal for the owners of the firm.

The infrastructures of many large companies today reflect the rigid engineering designs that have been with us since the Industrial Revolution. Those who moved forward faster, shunning legacy ideas of production, have been able to use infrastructure planning as a competitive advantage in itself. Toyota's ascendancy in automotive manufacturing is a good example of early identification of the introduction of flexibility in manufacturing, including technology, location, and product design. Similarly, Hewlett-Packard's application of "postponement"—the ability to avoid full assembly of all components until the product is ready to be delivered—required a rethinking of manufacturing and logistic infrastructure across the world. Companies have to almost religiously apply options thinking in designing and managing infrastructure in the modern world, fraught with disruptions caused by natural and human-made disasters. Planning, scheduling, and management—in both manufacturing and service industries—seem to depend a lot on historical data. Again, we see remnants of past experiences from the scale-and-scope economy at work. Arrival of computing power at almost zero cost allowed installation of large collection bins of historical data in the belief that such data will always allow better planning and management. In the last decade, enterprise resource management systems have proliferated, and data are piling in data warehouses at an exponential rate. It is not clear, however, how such data are used in reaching decisions and whether more historical data actually enhance planning and management. Attempts at precise forecasting of the future based on historical data are futile. A major role of historical data is providing information on uncertainty and risk and not on forecasting of future outcomes precisely.

In an era driven primarily by innovation, companies have to focus on incorporating uncertainty and flexibility in all aspects of business. These include

1. Designing an organization that provides flexibility to adapt and learn
2. Hiring, training, and retention policies to create human resources with maximum flexibility and diversity
3. Designing compensation and incentives that allow optimal management of uncertainty (rather than attempts at eliminating uncertainty)
4. Creating an infrastructure, both physical and financial, with maximum flexibility to cope with changing demand and needs

These conditions—flexible designs of organization, human resources, incentives, and infrastructure to cope with unavoidable (and not to be avoided) uncertainty—are necessary but not sufficient for success for companies in the innovation era. In addition, decision makers must make better decisions on how, when, and what to invest and how to maximize the value of their enterprises through holistic risk and portfolio management. They have to make decisions systematically based on market-based economic value and avoid the temptation to make *ad hoc* decisions. As you have seen in cases described in this book, decision makers can improve decision making through the application of systematic methodology, processes, and tools. A holistic framework that considers uncertainty and flexibility is essential for better management of today's and tomorrow's companies.

References

Baldwin, C. and K. Clark. 1997. Managing in an age of modularity. *Harvard Business Review* (September/October).

Berger, P.G., E. Ofek, and I. Swary. 1996 Investor valuation of the abandonment option. *Journal of Financial Economics* 42 (2).

Bernardo, Antonio E. and Bhagwan Chowdhry. 2002. Resources, real options, and corporate strategy. *Journal of Financial Economics*. 63 (2).

Black, Fischer and Myron Scholes. 1973. The Pricing of Options and Corporate Liabilities. *Journal of Political Economy* 81 (3).

Childs Paul, Steven Ott, and Alexander Triantis. 1998. Capital budgeting for inter-related projects: A real options approach. *Journal of Financial and Quantitative Analysis* 33 (3).

Cox, J.C. and S. A. Ross. 1976. The valuation of options for alternative stochastic processes. *Journal of Financial Economics* 3.

Fama, Eugene. 1964. Mandelbrot and the stable Paretian hypothesis. In *The Random Character of Stock Prices*. ed. Paul Cootner. Cambridge: The MIT Press. Originally published in *Journal of Business* 36 (4): 420–429.

Fama, Eugene. 1965. The Behavior of Stock Market Prices. *Journal of Business* 38 (1): 34–105.

Fama, Eugene. 1965. Random walks in stock market prices. *Financial Analysts Journal* 51 (1).

Fama, Eugene. 1968. Risk, return, and general equilibrium: Some clarifying comments. *Journal of Finance* 23 (1): 29–40.

Fama, Eugene. 1969. The adjustment of stock prices to new information. *International Economic Review* 10 (1): 1–21.

Fama, Eugene. 1970. Efficient capital markets: A review of theory and empirical work. *Journal of Finance* 25 (2): 383–417.

Fama, Eugene. 1971. Risk, return, and equilibrium. *Journal of Political Economy*. January/February.

Fama, Eugene and Arthur Laffer. 1971. Information and Capital Markets. *Journal of Business* (July).

Froot, K., D. Scharfstein, and J. Stein. 1994. A framework for risk management. *Viewpoint* 2 (November/December).

Gibson, R. and E. Schwartz . 1990. Stochastic convenience yield and the pricing of oil contingent claims. *Journal of Finance* 45 (3).

Hayes, R. and D. Garvin. 1982. Managing as if tomorrow mattered. *Harvard Business Review* 60 (3).

Kester, W. 1984. Today's options for tomorrow's growth. *Harvard Business Review* (March/April).

Luehrman, Timothy A. 1997. What's it worth? A general manager's guide to valuation. *Harvard Business Review* 75 (3).

Luehrman, Timothy A. 1998. Investment opportunities as real options: Getting started on the numbers. *Harvard Business Review* 76 (4).

Luehrman, Timothy A. 1998. Strategy as a portfolio of real options. *Harvard Business Review* 76 (5).

Majd, S. and R.S. Pindyck. 1987. Time to build, option value, and investment decision. *Journal of Financial Economics* 18 (March).

Malkiel, Burton G. 1987. Efficient market hypothesis. *The new palgrave: A dictionary of economics*. Vol. 2. eds. John Eatwell, Murray Milgate and Peter Newman. The Macmillan Press.

Mayers, David. 1998. Why firms issue convertible bonds: The matching of financial and real investment options. *Journal of Financial Economics* 47 (1).

McDonald, R. and D. Siegel. 1985. Option pricing when the underlying asset earns a below-equilibrium rate of return. *International Economic Review* 34 (1).

Merton, R. 1998. Applications of option-pricing theory: Twenty-five years later. *American Economic Review* (June).

Merton, Robert C. 1973. Theory of rational option pricing. *Bell Journal of Economics and Management Science* 4 (1).

Milne, Alistair and A. Elizabeth Whalley. 2000. Time to build, option value and investment decisions. *Journal Of Financial Economics* (56) 2.

Pindyck, R.S. 1993. Investments of uncertain cost. *Journal of Financial Economics* 34 (August).

Samuelson, Paul. 1972. Proof that properly anticipated prices fluctuate randomly. In *The Collected Scientific Papers of Paul A. Samuelson*. Vol. 3. ed. Robert C. Merton, chapter 198. Cambridge: The MIT Press. Originally published in *Industrial Management Review* 6 (2): 41–49.

Sharpe, William F. 1991. The arithmetic of active management. *The Financial Analysts' Journal*. 47 (1).

Smit, T. J. and Lenos Trigeorgis. 2004. *Strategic investment: Real options and games.* Princeton: Princeton University Press.

Trigeorgis, Lenos. 1996. *Real options: Managerial flexibility and strategy in resource allocation.* Cambridge: The MIT Press.

Tufano, P. 1996. Who manages risk? An empirical examination of risk management practices in the gold mining industry. *Journal of Finance*. 51 (4).

Index

For Product Safety Concerns and Information please contact our EU
representative GPSR@taylorandfrancis.com
Taylor & Francis Verlag GmbH, Kaufingerstraße 24, 80331 München, Germany